T0325148

Computational Topology for Biomedical Image and Data Analysis

Theory and Applications

CRC Press Focus Series in Medical Physics and Biomedical Engineering

Series Editors:
Magdalena S. Stoeva and Tae-Suk Suh

Recent books in the series:

Computational Topology for Biomedical Image and Data Analysis: Theory and Applications
Rodrigo Rojas Moraleda, Nektarios A. Valous, Wei Xiong, and Niels Halama

Computational Topology for Biomedical Image and Data Analysis

Theory and Applications

Rodrigo Rojas Moraleda
Nektarios A. Valous
Wei Xiong
Niels Halama

CRC Press
Taylor & Francis Group
Boca Raton London New York

CRC Press is an imprint of the
Taylor & Francis Group, an **informa** business

CRC Press
Taylor & Francis Group
6000 Broken Sound Parkway NW, Suite 300
Boca Raton, FL 33487-2742

© 2020 by Taylor & Francis Group, LLC
CRC Press is an imprint of Taylor & Francis Group, an Informa business

No claim to original U.S. Government works

Printed on acid-free paper

International Standard Book Number-13: 978-1-138-33634-6 (Hardback)

Visit the Taylor & Francis Web site at
http://www.taylorandfrancis.com

and the CRC Press Web site at
http://www.crcpress.com

with love to Emily Karx

Contents

SECTION II   Case studies

CHAPTER 3 ▪ Recognizing noise   87

CHAPTER 4 ▪ Image segmentation   93

CHAPTER 5 ▪ Point cloud characterization   103

Series Preface

CRC Press Focus Series in Medical Physics and Biomedical Engineering

Series Editors: Magdalena S. Stoeva and Tae-Suk Suh

About the Series

The *CRC Press Focus Series in Medical Physics and Biomedical Engineering* is a sub-series of the *CRC Press Series in Medical Physics and Biomedical Engineering*; a leading international book series and the official book series of the International Organization for Medical Physics (IOMP).

Titles in this sub-series explore hot topics targeting the applications of physical sciences, engineering, and mathematics in medicine and clinical research. They are available very quickly as eBooks and print-on-demand hardbacks, meeting the need for up-to-date texts in this rapidly developing field.

The *CRC Press Focus Series in Medical Physics and Biomedical Engineering* seeks (but is not restricted to) publications in the following topics:

- Artificial organs
- Assistive technology
- Bioinformatics
- Bioinstrumentation
- Biomaterials
- Biomechanics
- Biomedical engineering
- Clinical engineering

- Imaging

- Implants

- Medical computing and mathematics

- Medical/surgical devices

- Patient monitoring

- Physiological measurement

- Prosthetics

- Radiation protection, health physics, and dosimetry

- Regulatory issues

- Rehabilitation engineering

- Sports medicine

- Systems physiology

- Telemedicine

- Tissue engineering

- Treatment

The International Organization for Medical Physics

The International Organization for Medical Physics (IOMP) represents over 26,000 medical physicists worldwide and has a membership of 86 national member organizations, one affiliate, and six regional organizations, together with a number of corporate members. Individual medical physicists of all national member organizations are also automatically members.

The mission of IOMP is to advance medical physics practice worldwide by disseminating scientific and technical information, fostering the educational and professional development of medical physics, and promoting the highest quality medical physics services for patients.

A World Congress on Medical Physics and Biomedical Engineering is held every three years in cooperation with International Federation for Medical and Biological Engineering (IFMBE) and International Union for Physics and Engineering Sciences in

Medicine (IUPESM). A regionally based international conference, the International Conference of Medical Physics (ICMP), is held between world congresses. IOMP also endorses and supports international events - conferences, workshops, and courses.

The IOMP has several programmes to assist medical physicists in developing countries, such as the joint IOMP Library Programme and the Used Equipment Programme. The Travel Assistance Programme provides a limited number of grants to enable physicists to attend the world congresses.

The IOMP publishes, twice a year, an electronic bulletin, *Medical Physics World* and the *Medical Physics International Journal*. IOMP has an agreement with Taylor & Francis for the publication of the *Medical Physics and Biomedical Engineering series* and the Focus sub-series. IOMP members receive a discount.

IOMP collaborates with international organizations, such as the World Health Organisations (WHO), the International Atomic Energy Agency (IAEA), the International Union of Pure and Applied Physics (IUPAP), the International Science Council (ISC), and other international professional bodies, such as the International Radiation Protection Association (IRPA) and the International Commission on Radiological Protection (ICRP), to promote the development of medical physics and the safe use of radiation and medical devices.

Guidance on education, training, and professional development of medical physicists is issued by IOMP, which is collaborating with other professional organizations in development of a professional certification system for medical physicists that can be implemented on a global basis.

The IOMP website (www.iomp.org) contains information on all the activities and ways to interact with the IOMP.

Foreword

T HE monograph aims to provide an accessible yet rigorous introduction to topology and homology focused on a particular mathematical space, the *simplicial space*.

The first part is an introduction to *group theory* which provides convenient algebraic tools to operate and transform collections. Next, a mathematical construction: the *topological space* is presented based on properties of sets and subsets. The *simplicial topological space* is then presented using geometric and combinatorial formulations. Considering all these properties and constructions, a discrete algebraic invariant called the *homology group* is then introduced for comparing and classifying topological spaces.

The main contribution of this work to the literature of computational homology is to provide a compact pipeline from the foundations of topology to the practicalities of simplicial homology, with straightforward definitions for simplicial complexes in three ways: geometric oriented and unoriented, as well combinatorial.

Preface

Topology is likely the newest branch of mathematics compared to fields such as geometry, number theory, or algebra which can be traced back to antiquity. Topology started in the seventh century under the name *Analysis Situs* (the analysis of position) and is replaced later in the 19th century by *topology* [43]. According to the *Oxford English Dictionary*, the word "topology" is derived from the Greek words "topos" ($\tau \acute{o} \pi o \varsigma$), meaning place, and "-logy" ($\lambda o \gamma \iota \alpha$), a variant of the verb $\lambda \acute{\varepsilon} \gamma \varepsilon \iota \nu$, meaning to speak. As such, topology speaks about places: how local neighborhoods connect to each other to form a space.

A general definition of topology is the following: *topology concerns the study of the way in which constituent parts are interrelated or arranged*. This definition fits with the use of the term in a wide range of applications, e.g., computer networks, biology, communications, etc. In mathematics the term is more precise: *topology studies properties of spaces that are invariant under any continuous deformation*. Hence, the spaces considered in topology have the ability to keep their fundamental properties despite stretching and contracting like rubber and without breaking or tearing. A practical approach for studying such spaces is provided by *algebraic topology*.

Contributors

Anna Berthel
National Center for Tumor
   Diseases
Heidelberg, Germany

Katja Breitkopf-Heinlein
Medical Faculty Mannheim
Heidelberg University
Germany

Pornpimol Charoentong
National Center for Tumor
   Diseases
Heidelberg, Germany

Steven Dooley
Medical Faculty Mannheim
Heidelberg University
Germany

Dieter W. Heermann
Institute for Theoretical
   Physics
Heidelberg University
Germany

Jeremy Mann
Department of Mathematics
University of Notre Dame
Notre Dame, IN, USA

Luis Salinas
Department of Informatics
Federico Santa María Technical
   University
Valparaíso, Chile

Meggy Suarez-Carmona
National Center for Tumor
   Diseases
Heidelberg, Germany

Inka Zörnig
National Center for Tumor
   Diseases
Heidelberg, Germany

Authors

Rodrigo Rojas Moraleda obtained his PhD in computer science from Federico Santa María Technical University, Chile. He has research experience in the fields of computational biomedicine and applied mathematics. His current affiliation is with the German Cancer Research Center (DKFZ) and the National Center for Tumor Diseases (NCT), Heidelberg, Germany.

Nektarios A. Valous obtained his PhD in biosystems engineering from University College Dublin, Ireland. He has research experience in the fields of computational biomedicine and interdisciplinary physics. His current affiliation is with the German Cancer Research Center (DKFZ) and NCT, Heidelberg, Germany.

Wei Xiong obtained her PhD in mathematics from Heidelberg University, Germany. She has worked in the statistical physics and theoretical biophysics group (Institute for Theoretical Physics) and has research experience in the fields of computational mathematics and scientific computing. Her current affiliation is with the Institute for Theoretical Physics at Heidelberg University, Germany.

Niels Halama is the head of the department of translational immunotherapy at the German Cancer Research Center, Germany. He has research experience in the fields of cancer immunology and computational biomedicine. His current affiliation is with the German Cancer Research Center (DKFZ) and NCT, Heidelberg, Germany.

I

Theoretical foundations

1

Theoretical Foundations

Elements of topology and homology

L IKE all branches of mathematics, algebraic topology and homology have a particular lexicon. This chapter intends to immerse the reader in the essential techniques and vocabulary of group theory, topology, and homology as well as providing illustrative examples for an accessible yet rigorous introduction to these topics. In this regard, the essential components of the theory required to present homology are introduced in an accessible way. This chapter does not aim to be a mathematical compendium.

The first part of the chapter presents a terse introduction to group theory which provides the necessary algebraic tools to operate and transform collections. Then, a topological space is defined based on the properties of collections and their subsets. In section 1.5.1, the simplicial topological space is introduced using geometric and combinatorial formulations. Finally, section 1.6 presents a discrete algebraic invariant: the homology group which is used to compare and classify simplicial topological spaces.

Readers that are interested in a more formal and rigorous mathematical discussion should consult the literature proposed at the end of the chapter.

1.1 ELEMENTS OF GROUP THEORY

Group theory is introduced using set theory as the canvas. Set theory is an influential and engaging subject which (in this monograph) will not be explored in its entirety; however, it provides a more natural and intuitive way to present the essentials of group theory.

1.1.1 Set

Intuitively, a set is a collection of "things". A method to construct a set is the following: identify a domain, specify an arbitrary rule, and then gather together these "things" in the domain that satisfy the rule. As an example consider integer numbers; a variety of rules can be defined in order to gather integers into different sets:

- an infinite set of even numbers: $\{\ldots, -4, -2, 0, 2, 4, \ldots\}$

- an infinite set of prime numbers: $\{2, 3, 5, 7, 11, 13, 17, \ldots\}$

An early attempt for a rigorous definition of sets was given by Zermelo's paper (1908) titled *Untersuchungen über die Grundlagen der Mengenlehre. I* [54] which became a milestone for the modern theory of sets. Here is the definition:

Definition 1.1 (Naive set theory) *A set is a collection of definite, distinguishable objects of perception or thought conceived as a whole (single and abstract object).*

However, Zermelo's intuitive approach to set theory quickly becomes paradoxical such as in Russell's antinomy. These types of problems have their origin in the fact that terms commonly used in mathematics such as "set", "number", "function", "sequence", "statement", "formula", and others are notions that originate in ordinary language with rational meaning in everyday use, but cannot be fully defined in a mathematically rigorous way [52]. For example:

a) $A = \{\text{smallest prime} > 20^{13^{15^{17^{19^{21^{23^{25}}}}}}}\}$

b) $A = \{\text{the set of all sets with exactly one element}\}$

c) $R = \{ x \mid x \notin x \}$, then $R \in R \iff R \notin R$

These are sets that cannot be specified but only described and mathematics does not have a consistent theory to deal with them. To avoid such ambiguities, a modern approach to set theory is given by axiomatic set theory. In axiomatic set theory, an object is specified through a predicate, that is regarding a property (or properties) that it possesses [20]. The approach given by [54] is called naive set theory as opposed to the more rigorous axiomatic set theory. Although naive set theory is not always rigorous, all results in this work can be derived from it.

Definition 1.2 (Cardinality) *The cardinality of a set A denoted by* $|A|$, $\#A$, $card(A)$, $\overline{\overline{A}}$ *or* $n(A)$ *is the number of elements in A.*

Definition 1.3 (Subsets) *A is a subset of B denoted by* $A \subseteq B$, *if and only if (iff) every element of A is in B.*

The relationship $A \subseteq B$ is also known as *inclusion* or *containment*. The inverse: A is not a subset of B is denoted by $A \nsubseteq B$.

Properties:

- $A \subseteq B$ iff their intersection is equal to A;
  $A \subseteq B \Leftrightarrow A \cap B = A$

- $A \subseteq B$ iff their union is equal to B;
  $A \subseteq B \Leftrightarrow A \cup B = B$

- A finite set A is a subset of B iff the cardinality of their intersection is equal to the cardinality of A;
  $A \subseteq B \Leftrightarrow |A \cap B| = |A|$

The definition of subsets runs into deductions like $A \subseteq A$. A *proper subset* on the other hand is a subset that is strictly not the same as the original set.

Definition 1.4 (Proper subsets) *A is a proper subset of B denoted by* $A \subset B$, *iff every element in A is also in B and at least one element in B is not in A.*

The inverse relationship: A being not a proper subset of B is denoted by $A \not\subset B$.

Definition 1.5 (Null set) *A set with no elements is denoted by* $\emptyset$.

Properties:

- The empty set is a subset of every set;
  $A = \{\heartsuit, \diamond, \clubsuit, \spadesuit\}$; $\emptyset \subset A$

- The empty set is a subset of the empty set itself; $\emptyset \subseteq \emptyset$

Definition 1.6 (Power set) *Given a set S, the power set of S denoted by $\mathcal{P}(S)$ is the set of all subsets of S including the empty set and S itself.*

If S is a finite set with $|S| = n$ then the number of subsets of S is $|\mathcal{P}(S)| = 2^n$.

Example: The set $S = \{a, b, c\}$; $|S| = 3$; $\mathcal{P}(S) = \{\emptyset, \{a\}, \{b\}, \{c\}, \{a, b\}, \{a, c\}, \{b, c\}, \{a, b, c\}\}$; $|\mathcal{P}(S)| = 2^3 = 8$ elements.

Definition 1.7 (Binary operator) *A binary operator symbolized as a placeholder $*$ on a non-empty set S is a rule that assigns to each ordered pair of elements (a,b) in S, a unique element in S:*

$$* : S \times S \to S$$

Examples include addition $(a + b)$, multiplication $(a \times b)$, subtraction $(a - b)$, exponentiation a^b (on $\mathbb{R}_+$), the symmetric difference of sets $(A \oplus B)$, etc.

Definition 1.8 (Partial order) *A (non-strict) partial order is a binary relation denoted by $\leq$ over a set S that satisfies the following axioms; for all $\heartsuit, \heartsuit$, and $\clubsuit$ in S:*

- *Reflexivity: every element is related to itself;*
  $\heartsuit \leq \heartsuit$

- *Antisymmetry: two distinct elements cannot be mutually related;*
  if $\heartsuit \leq \heartsuit$ and $\heartsuit \leq \heartsuit$ then $\heartsuit = \heartsuit$

- *Transitivity;*
  if $\heartsuit \leq \clubsuit$ and $\clubsuit \leq \heartsuit$, then $\heartsuit \leq \heartsuit$

Definition 1.9 (Partial order set: poset) *A set equipped with a partial order relation is called a partially ordered set or a poset.*

1.1.2 Group

Definition 1.10 (Group) *A group is an algebraic structure consisting of a set of elements denoted by G together with a binary operator symbolized by* ∗. *The pair is symbolized by* $(G, *)$. *To qualify as a group, the pair* $(G, *)$ *must satisfy four requirements known as* **group axioms:**

- *The group has an identity element often symbolized as e;*
  $\exists\, e \in G : \forall\, a \in G : e * a = a = a * e$

- *Each element has an inverse;*
  $\forall\, a \in G : \exists\, b \in G : a * b = e = b * a$

- *The group is closed under the operator* ∗;
  $\forall\, a, b \in G : a * b \in G$

- *Associativity;*
  $\forall\, a, b, c \in G : a * (b * c) = (a * b) * c$

Example: Groups under addition $(G, +)$:

- $\mathbb{R}$: set of real numbers

- $\mathbb{Q}$: set of rational numbers

- $\mathbb{Z}$: set of integer numbers

- $\mathbb{Z}_2$: set of integer elements $\{0, 1\}$

Example: Groups under multiplication $(G, \times)$:

- $\mathbb{R} \setminus \{0\}$: set of real numbers not including zero

- $\mathbb{Q} \setminus \{0\}$: set of rational numbers

- $\mathbb{Z}_2 \setminus \{0\} = \{1\}$: the operator $\times$ defined by:

$\times$	0	1
0	0	1
1	1	0

Groups of numbers with multiplicative structure cannot include 0 because 0 does not have a multiplicative inverse. Similarly, $\mathbb{Z} \setminus \{0\}$ is not a group under multiplication because given an $x \neq 1 \in \mathbb{Z}$, a multiplicative inverse $x^{-1} \in \mathbb{Z}$ does not exist such that $x \times x^{-1} = 1$ where 1 is the identity. For example, for $x = 3$:

$$x \times x^{-1} = 1 \;\rightarrow\; x^{-1} = \tfrac{1}{3} \text{ but } \tfrac{1}{3} \notin \mathbb{Z}$$

Example: $G = \{0\}$ under addition is the group $(G, +)$:

- Identity: 0 added to anything else in the group is 0 $(0+0 = 0)$, 0 is the identity

- Inverse: 0 is the additive inverse of 0

- Associativity: since the only element is 0, associativity is trivially satisfied

- Closure: since the only element is 0, $0 + 0$ is in the group

Example: $G = \{-1, 1\}$ under multiplication is the group $(G, \times)$:

- operator $\times$ is defined by:

$\times$	-1	1
-1	1	-1
1	-1	1

- Identity: element 1 satisfies $1 \times -1 = -1$ and $-1 \times 1 = -1$

- Inverse: element -1 satisfies $1 \times -1 = -1$ and $-1 \times 1 = -1$

- Associativity: any multiplication of two elements satisfies $a \times (b \times c) = (a \times b) \times c$

- Closure: any combination of multiplication between two elements is an element of the group and the set, thus the operator is closed

Definition 1.11 (Subgroup) *Given a group $(G, *)$ and a subset of G H ($H \subseteq G$), then $(H, *)$ is called a subgroup of $(G, *)$ iff $(H, *)$ is a group.*

Example:

- $(\mathbb{R}_+, \times)$ is a subgroup of $(\mathbb{R} \setminus \{0\}, \times)$ with identity 1

- $(\mathbb{R}_-, \times)$ is not a subgroup of $(\mathbb{R} \setminus \{0\}, \times)$ because it is not closed: $(-1) \times (-1) = 1$, $1 \notin \mathbb{R}_-$ (there is no identity)

Definition 1.12 (Abelian group) *An Abelian group also known as a commutative group is a group $(G, *)$ that satisfies the axiom of commutativity[1]:*

---

[1] **Convention:** an Abelian group without an operator is a group under addition; therefore, expressions of the kind "given an Abelian group G" should be understood as "given an Abelian group $(G, +)$."

- *Commutativity;*
  $\forall a, b \in G : a * b = b * a$

Definition 1.13 (Non-Abelian group) *A non-Abelian group is a group* $(G, *)$ *in which there are at least two elements* $a, b \in G$ *such that* $a * b \neq b * a$.

Matrix groups under multiplication are non-commutative or non-Abelian groups.

Definition 1.14 (Finitely generated group) *Given a group* $(G, *)$ *and a subset* $S \subset G$, *the finitely generated group S denoted by* $\langle S \rangle$ *is a subgroup consisting of all elements of G that can be expressed as finite combinations (under the operator* $*$ *) of points in S and their inverses.*

Properties of $\langle S \rangle$:

- $\langle S \rangle$ *is the smallest subgroup of G containing every element of S*

- *If* $G = \langle S \rangle$ *then the elements of S are called generators or group generators*

- *If* $S = \emptyset$ *then* $\langle S \rangle$ *is the trivial group e which is a group consisting of a single element (identity)*

Definition 1.15 (Rank of a group) *The rank of a group G is the cardinality of the smallest subset generating G:*

$$rank(G) = \{|X| : X \subset G, \langle X \rangle = G\}$$

Example:

$rank(\mathbb{Z}) = 1$, because $\mathbb{Z} = \langle \{1\} \rangle$

$rank(\mathbb{Z} \times \mathbb{Z}) = 2$, because $\mathbb{Z} \times \mathbb{Z} = \mathbb{Z}^2 = \langle \{(0,1), (1,0)\} \rangle$

Definition 1.16 (Free Abelian group) *Consider a set* $X = \{x_1, x_2, \ldots, x_k\}$, *then a free Abelian additive group G generated by the elements of X consist of all elements of the form:*

$$n_1 x_1 + n_2 x_2 + \ldots + n_k x_k = \sum_{i=1}^{k} n_i x_i$$

where $n_i \in \mathbb{Z}$. *The integer combination of elements of X is denoted by* $\mathbb{Z}[X]$. *The addition of elements of G works as follows:*
$(n_1 x_1 + n_2 x_2 + \ldots + n_k x_k) + (m_1 x_1 + m_2 x_2 + \ldots + m_k x_k) = (n_1 + m_1)x_1 + (n_2 + m_2)x_2 + \ldots + (n_k + m_k)x_k.$

The set X acts as a basis of G hence $rank(G) \leq k$, but this is not universally true because some of the elements of X may be generated by other elements of the same set.

Example: A free Abelian additive group generated by $\{\clubsuit, \spadesuit\}$ is $\mathbb{Z}[\clubsuit, \spadesuit]$ where $\mathbb{Z}\clubsuit + \mathbb{Z}\spadesuit = \{\clubsuit m + \spadesuit n \mid m, n \in \mathbb{Z}\}$ may contain elements like:

$$\{19\clubsuit + 73\spadesuit\}, \{\clubsuit + 8\spadesuit\}, \{-4\clubsuit\}, \{m\clubsuit + n\spadesuit\}$$

1.1.3 Ring and field

In the hierarchy of mathematical objects two algebraic structures emerge: rings and fields. Rings refer to sets equipped with the operators of addition and multiplication bonded together by a distributive property. The focus here is mainly on a more constrained structure: the field.

Definition 1.17 (Ring) *A set $\mathcal{R}$ equipped with two binary operators, namely addition (+) and multiplication ($\times$) is called a ring $(\mathcal{R}, +, \times)$ if the following collections of axioms called ring axioms are satisfied:*

$\mathcal{R}$ *is an Abelian group under addition:*

- $(a + b) + c = a + (b + c)$, $\forall a, b, c \in \mathcal{R} \Rightarrow$ (+ *is associative*)
- $a + b = b + a$, $\forall a, b \in \mathcal{R} \Rightarrow$ (+ *is commutative*)
- *an element $0_+ \in \mathcal{R}$ such that $a + 0_+ = a$, $\forall a \in \mathcal{R} \Rightarrow 0_+$ (additive identity)*
- *for each $a \in \mathcal{R}$ exists $-a \in \mathcal{R}$ such that $a + -a = 0 \Rightarrow -a$ (additive inverse of a)*

The operator $\times$ is associative therefore closed:

- $(a \times b) \times c = a \times (b \times c)$, $\forall a, b, c \in \mathcal{R} \Rightarrow \times$ *is associative*

Multiplication is distributive with respect to addition:

- $a \times (b + c) = (a \times b) + (a \times c)$, $\forall a, b, c \in \mathcal{R} \Rightarrow$ *left distributivity*
- $(b + c) \times a = (b \times a) + (c \times a)$, $\forall a, b, c \in \mathcal{R} \Rightarrow$ *right distributivity*

Definition 1.18 (Commutative ring) *A ring $(\mathcal{R}, +, \times)$ is commutative if the multiplication operator $\times$ is commutative:*

$$\forall a, b \in \mathcal{R}, \quad a \times b = b \times a$$

Definition 1.19 (Ring with unity) *A ring* $(\mathcal{R}, +, \times)$ *is called ring with unity if an element exists in* $\mathcal{R}$ *that is different from* 0_+. *The multiplication identity is denoted by* $1_\times$:

$$\exists \, 1_\times \in \mathcal{R} \text{ such that } a \times 1_\times = a, \forall a \in \mathcal{R}$$

Definition 1.20 (Field) *A set* F *is a field with respect to addition and multiplication if it satisfies the following:*

- *F is an Abelian group under addition*

- *$F \setminus \{0\}$ is an Abelian group under multiplication*

- *the multiplication operator* $(\times)$ *distributes across addition:* $\forall \, a, b, c \in F : a \times (b + c) = a \times b + a \times c$

Example: Some examples of fields are:

- $\mathbb{R}$: set of real numbers

- $\mathbb{Q}$: set of rational numbers

- $\mathbb{Z}_2$: set $\{0, 1\}$ with:

+	0	1		×	0	1
0	0	1		0	0	0
1	1	0		1	0	1

Definition 1.21 (Formal linear combination) *A formal linear combination of elements of a set* S *with respect to a given field, e.g., the field* $\mathbb{Z}$ *of integers is a function:*

$$F : S \to \mathbb{Z}$$

such that given $s \in S$ *only a finite number of values are non-zero* $(F(s) \neq 0)$.

Identifying each element $x \in S$ with a function that takes the value 1 on x and zero on all other elements of S, then any element $F \in \mathbb{Z}\langle S \rangle$ can be written uniquely in the form $F = \sum_{i=1}^{m} a_i x_i$ where $x_1, \ldots, x_m$ are the elements of S for which $F(x_i) \neq 0$ and $a_i = F(x_i)$. Thus, S is a basis for $\langle S \rangle$ which is finite-dimensional iff S is a finite set. Later, *formal linear combinations* are going to be used in order to construct a special topological space where digital images can be analyzed using homology.

Definition 1.22 (Free vector space) *A free vector space over a field F generated by elements $X = \{x_1, x_2, \ldots, x_k\}$ consists of all elements of the form:*

$$n_1 x_1 + n_2 x_2 + \ldots + n_k x_k \quad k \in \mathbb{N}$$

where $n_i \in F$. Note that this is a space of formal linear combinations of elements of X with $n_i \in \mathbb{Z}$.

Example: A free vector space over the field $\mathbb{Z}_2$ generated by elements $X = \{x_1, x_2\}$ has the form:

$$n_1 x_1 + n_2 x_2$$

where $n_1, n_2 \in \{0, 1\}$. A limited number of combinations remains for the elements:

$$
\begin{aligned}
n_1 = 0, n_2 = 0 &\implies 0x_1 + 0x_2 = 0 \\
n_1 = 1, n_2 = 0 &\implies 1x_1 + 0x_2 = x_1 \\
n_1 = 0, n_2 = 1 &\implies 0x_1 + 1x_2 = x_2 \\
n_1 = 1, n_2 = 1 &\implies 1x_1 + 1x_2 = x_1 + x_2
\end{aligned}
$$

$$\bullet \qquad \bullet \qquad \bullet$$

$$\mathbb{Z}_2[x_1, x_2] = \{0, \ x_1, \ x_2, \ x_1 + x_2\}$$

Definition 1.23 (Coset) *Given a subgroup H of an Abelian group G then for any $a \in G$ the induced set $a * H$ is the following:*

$$a * H = \{a * h \mid h \in H\}$$

This is called the coset a of H denoted by $H(a)$.

Example of cosets generated by points $a, b, c, d, e, f, j \in G$ over the subgroup H are shown in Fig. 1.1i.

Example: Let $G = \mathbb{R} \setminus \{0\}$, $H = \mathbb{R}_+$, and $* = \times$, then:

- $\pi \times \mathbb{R}_+$ is a coset which is the same as $\mathbb{R}_+$

- $-1 \times \mathbb{R}_+ = \mathbb{R}_-$ is a coset but not in the same subgroup H

Definition 1.24 (Normal subgroup) *A subgroup $(H, *)$ of a group $(G, *)$ is normal in G if $g * H = H * g$ for all $g \in G$. In a normal subgroup of a group G the right and left cosets are the same.*

Example:

- $(\mathbb{R}_+, \times)$ is a normal subgroup of $(\mathbb{R} \setminus \{0\}, \times)$

- $(\mathbb{R}_-, \times)$ is not a normal subgroup of $(\mathbb{R} \setminus \{0\}, \times)$ because it is not a group (not closed $-1 \times -1 = 1$)

1.1.4 Homomorphism

A homomorphism is a structure-preserving function in the domains of algebra, discrete mathematics, groups, rings, graphs, and lattices. Structure-preserving functions between groups preserve their natural (algebraic) structure.

Definition 1.25 (Homomorphism) *Given groups $(G, \star)$ and $(G', *)$ then a function $\phi : G \to G'$ is a homomorphism if it satisfies:*

$$\phi(a \star b) = \phi(a) * \phi(b) \quad \forall a, b \in G$$

Notation: $Hom(G, G')$ denotes the set of all homomorphisms from G to G'. The $Hom(G, G)$ from G to G is called an automorphism and is denoted by $Hom(G)$.

Example: Given the multiplicative group of positive real numbers $(\mathbb{R}_{>0}, \times)$ and the additive group of real numbers $(\mathbb{R}, +)$, then the mapping:

$$\log_{10} : \quad (\mathbb{R}_{>0}, \times) \quad \to \quad (\mathbb{R}, +)$$
$$x \quad \to \quad \log_{10}(x)$$

satisfies the group homomorphism properties:

$$\forall \, x, y \in \mathbb{R}_{>0} : \log_{10}(x \times y) = \log_{10}(a) + \log_{10}(b)$$

in consequence, $\log_{10}$ is a group homomorphism.

Example: Let $\phi : \mathbb{Z}^+ \to \mathbb{Z}^2$ be a function such that ($x \in \mathbb{Z}^+$):

$$\phi(x) = \begin{cases} 0 & \text{if } x \text{ is even} \\ 1 & \text{if } x \text{ is odd} \end{cases}$$

then ϕ is a homomorphism. If $(x + y)$ is even then x and y are even numbers or x and y are odd numbers. If x and y are even numbers with $\phi(x) = \phi(y) = 0$, then $\phi(x) + \phi(y) = 0 + 0 = 0$ in $\mathbb{Z}_2$. If x and y are odd numbers with $\phi(x) = \phi(y) = 1$, then $\phi(x) + \phi(y) = 1 + 1 = 0$ in $\mathbb{Z}_2$:

$$\phi(x + y) = 0 = \phi(x) + \phi(y) \quad (\mathbb{Z}_2)$$

If $(x + y)$ is odd then at most one of them x or y is odd. If x is even then $\phi(x) = 0$ but if y is odd then $\phi(y) = 1$, therefore $\phi(x) + \phi(y) = 0 + 1 = 1$ in $\mathbb{Z}_2$:

$$\phi(x + y) = 1 = \phi(x) + \phi(y) \quad (\mathbb{Z}_2)$$

Definition 1.26 (Isomorphism) *Let a function* $\phi : G \to H$ *be a homomorphism. If* ϕ *has also a one-to-one correspondence then* ϕ *is called an isomorphism. Two groups G and H are called isomorphic and are denoted by:*

$$G \cong H$$

if an isomorphism exists between them.

Definition 1.27 (Kernel of homomorphism) *Given groups* $(G, \star)$ *and* $(G', *)$ *and a homomorphism* $\phi : G \to G'$, *then the kernel of the homomorphism is given by:*

$$ker\phi = \{a \in G \mid \phi(a) = e'\}$$

where e' *is the identity in* G' *(Fig. 1.1ii).*

Definition 1.28 (Quotient group) *Given a normal subgroup* $(H, *)$ *of a group* $(G, *)$ *(def. 1.24) then the cosets of H in G form a group* G/H *under the operator* $*$ *such that:*

$$(a * H)(b * H) = (a * b) * H \; \forall a, b \in G \qquad (1.1)$$

This group is called the factor or quotient group of G and H.

Example: Consider a group G and a subgroup $ker\phi$. The schematic illustrates their quotient group denoted by $\frac{G}{ker\phi}$. The diagram shows four cosets generated by four arbitrary points $a, b, c, d \in G$. The union of all cosets generated by all points in G over the subgroup $ker\phi$ constitutes $G/ker\phi$ (Fig. 1.1iii).

Example: Consider the normal subgroup $3\mathbb{Z}$ of $(\mathbb{Z}, +)$:

$$3\mathbb{Z} = \{\ldots -6, -3, 0, 3, 6, \ldots\}$$

The cosets of $3\mathbb{Z}$ in $\mathbb{Z}$ are:

Coset of $0 \in \mathbb{Z}$: $\quad [0] = 0 + 3\mathbb{Z} = \{\ldots, -6-3, 0, 3, 6, \ldots\} = 3\mathbb{Z}$

Coset of $1 \in \mathbb{Z}$: $\quad [1] = 1 + 3\mathbb{Z} = \{\ldots, -5-2, 1, 4, 7, \ldots\} = 3\mathbb{Z}$

Coset of $2 \in \mathbb{Z}$: $\quad [2] = 2 + 3\mathbb{Z} = \{\ldots, -4-1, 2, 5, 8, \ldots\} = 3\mathbb{Z}$

$\Rightarrow [0] \cap [1] \cap [2] = \emptyset \quad$ and $\quad [0] \cup [1] \cup [2] = \mathbb{Z}$

$\Rightarrow \mathbb{Z}/3\mathbb{Z} = \{[0], [1], [2]\}$

The group $\mathbb{Z}/3\mathbb{Z}$ is given by the addition operator:

+	[0]	[1]	[2]
[0]	[0]	[1]	[2]
[1]	[1]	[2]	[0]
[2]	[2]	[0]	[1]

The subgroup $n\mathbb{Z}$ of $\mathbb{Z}$ is normal. The cosets of $\mathbb{Z}/n\mathbb{Z}$ are:

$n\mathbb{Z}$

$1 + n\mathbb{Z}$

$2 + n\mathbb{Z}$

$\vdots$

$(n-1) + n\mathbb{Z}$

1.1.5 Modular arithmetic

Given $a, b \in \mathbb{Z}$ and $n \neq 0$, a unique q and r can be found such that:

$$a = qb + r \quad \text{where } 0 \le r < b$$

For $b = 5$ and $a := \{0 \ldots 14\}$:

$0 = 0 * 5 + 0$	$5 = 1 * 5 + 0$	$10 = 2 * 5 + 0$
$1 = 0 * 5 + 1$	$6 = 1 * 5 + 1$	$11 = 2 * 5 + 1$
$2 = 0 * 5 + 2$	$7 = 1 * 5 + 2$	$12 = 2 * 5 + 2$
$3 = 0 * 5 + 3$	$8 = 1 * 5 + 3$	$13 = 2 * 5 + 3$
$4 = 0 * 5 + 4$	$9 = 1 * 5 + 4$	$14 = 2 * 5 + 4$

A pattern is observed in the values of r (remainder); in this example r grows from 0 to $b - 1$. The rules followed by the remainders are called modular arithmetic.

Definition 1.29 (Congruence modulo m) *Given two integers a and b, then a and b are **congruent modulo** m, if $a - b$ is a multiple of m; the variable m is known as **modulus** and is denoted by $a \equiv b \,(mod\, m)$.*

Example: $42 \equiv 27 \,(mod\, 5)$ $42 - 27 = 15 = 3 * 5$

In relation to the remainders, two integers a and b are **congruent modulo** m if both share the same remainder when divided by m.

Example: $42 = 8*5 + 2$ and $27 = 5*5 + 2$ $\Rightarrow 42 \equiv 27 \,(mod\,5)$

Properties:
Given $a, b, c, d, k \in \mathbb{Z}$, $a \equiv b \,(mod\,m)$, and $c \equiv d \,(mod\,m)$ then the following axioms are satisfied:

$$ka \equiv kb \,(mod\,m)$$

$$a + c \equiv b + d \,(mod\,m)$$

$$a * c \equiv b * d \,(mod\,m)$$

$$a^n \equiv b^n \,(mod\,m) \quad (n \in \mathbb{Z}_{\geq 0})$$

Definition 1.30 (Congruence classes) *Congruence modulo m satisfies the axioms of an* **equivalence relation** *denoted by R:*

Reflexive: $\forall x \in A, (x, x) \in R$

Symmetric: $\forall x, y \in A, (x, y) \in R \Rightarrow (y, x) \in R$

Transitive: $\forall x, y, z \in A, (x, y) \in R \wedge (y, z) \in R \Rightarrow (x, z) \in R$

Congruence modulo m is an **equivalence relation** *partitioning the set of integer numbers into exactly m distinct* **congruence classes**. *The notation for the congruence class of an integer a modulo m is the following:*

$$[a]_m = \{b \in \mathbb{Z} \mid a \equiv b(mod\,m)\}$$

$[a]_m$ *is the set of integers that have the same remainder with a when divided by a multiple of m.*

Example: Considering the congruence modulo 2 then the integers under this relation are:

$$[0]_2 = \{\ldots, -4, -2, 0, 2, 4, \ldots\} \qquad \text{even numbers}$$
$$[1]_2 = \{\ldots, -3, -1, 1, 3, 5, \ldots\} \qquad \text{odd numbers}$$
$$[2]_2 = \{\ldots, -4, -2, 0, 2, 4, \ldots\} = [0]$$
$$[3]_2 = \{\ldots, -3, -1, 1, 3, 5, \ldots\} = [1]$$
$$\vdots \qquad\qquad \vdots \qquad\qquad \vdots$$

Because $\mathbb{Z}$ is an infinite set, the number of congruence classes is also infinite but it is possible to represent all congruence classes using the smallest non-negative members known as **principal representatives** (in this example 0 and 1). The m congruence classes induced by the relation $\equiv$ $(mod\ m)$ form a set of numbers called the **residue system mod** m denoted by $\mathbb{Z}_m$:

$$\mathbb{Z}_m = \{[0]_m, [1]_m, [2]_m, \ldots, [m-1]_m\}$$

To simplify things, the notation of $\mathbb{Z}_m$ is the following:

$$\mathbb{Z}_m = \{0, 1, 2, \ldots, m-1\}$$

The residue system satisfies:

Addition: $[a]_m + [b]_m = [a+b]_m$

Multiplication: $[a]_m \times [b]_m = [a \times b]_m$

Example: Addition and multiplication tables on $\mathbb{Z}_5$:

+	0	1	2	3	4
0	0	1	2	3	4
1	1	2	3	4	0
2	2	3	4	0	1
3	3	4	0	1	2
4	4	0	1	2	3

×	0	1	2	3	4
0	0	0	0	0	0
1	0	1	2	3	4
2	0	2	4	1	3
3	0	3	1	4	2
4	0	4	3	2	1

In the multiplicative table of $\mathbb{Z}_5$ every number except 0 has an inverse; that is for each number there is another one that when multiplied gives the multiplicative inverse 1. In contrast, consider the multiplicative table of $\mathbb{Z}_6$:

×	0	1	2	3	4	5
0	0	0	0	0	0	0
1	0	1	2	3	4	5
2	0	2	4	0	2	4
3	0	3	0	3	0	3
4	0	4	2	0	4	2
5	0	5	4	3	2	1

Only numbers 1 and 5 have a multiplicative inverse thus showing that the multiplicative inverse is not guaranteed. In general, given a prime number p then every non-zero element on $\mathbb{Z}_p$ has a multiplicative inverse.

Few words about the Cartesian product

The Cartesian product of two sets A and B is the set of all possible ordered pairs (a,b) from A and B denoted by the operator $\times$:

$$A \times B = \{(a,b) \mid a \in A \land b \in B\}$$

Example: Given two sets:

$$
\begin{aligned}
A &= \{1,2,3\} \\
B &= \{x,y,z\} \\
A \times B &= \begin{pmatrix} (1,x) & (1,y) & (1,z) \\ (2,x) & (2,y) & (2,z) \\ (3,x) & (3,y) & 3,z) \end{pmatrix}
\end{aligned}
$$

The two sets in the product can also be the same set, e.g., set of integer numbers $\mathbb{Z}$. The set of all ordered pairs of integers is the Cartesian product $\mathbb{Z} \times \mathbb{Z}$ which is denoted by $\mathbb{Z}^2$ also known as a point lattice. The n-ary Cartesian product is the set of all possible n-tuples from n different sets $A \times B \times C \ldots$ thus the n-ary Cartesian product of one set with itself is expressed as $A \times A \times A \times \cdots = A^n$.

1.2 TOPOLOGICAL SPACES

The aim of this section is to formalize the notion of a *topological space*. Topological spaces are the canvas for performing image and data analysis in this work.

Definition 1.31 (Space) *The use of the term space refers to a set of points with some added structure [25].*

Definition 1.32 (Topology axioms) *Let X be a set and $\mathcal{T}$ be a family of subsets of X, then $\mathcal{T}$ is called a topology on X if:*

1. *The empty set $\emptyset \in \mathcal{T}$ and $X \in \mathcal{T}$*

2. *Arbitrary (finite or infinite) unions of elements of $\mathcal{T}$ belong to $\mathcal{T}$ $U \subseteq \mathcal{T} \implies \bigcup_{A \in U} A \in \mathcal{T}$*

3. *Any finite intersection of elements of $\mathcal{T}$ is in $\mathcal{T}$ $U_1, \ldots, U_n \in \mathcal{T} \implies \bigcap_{k=1}^{n} U_k \in \mathcal{T}$*

The elements of a topology $(X, \mathcal{T})$ defined on a set X are subsets of X. Then, for any non-empty set X at least two topologies can be defined:

1. The **trivial topology** $\mathcal{T}_t = \{\emptyset, X\}$ containing no other set than $\emptyset$ and X

2. The **discrete topology** $\mathcal{T}_d = \mathcal{P}(X)$ of all subsets of X (power set of X) where every subset is an open set

Definition 1.33 (Open set) *Any subset of* $(X, \mathcal{T})$ *is called an* **open set**.

Definition 1.34 (Closed set) *The complement of an open set is called a* **closed set***: for any* $U \in T$ *its complement* $F = X \setminus U$ *is closed.*

Fig. 1.1v shows a) a closed set C_2 defined in an open set O_1 and b) an open set O_2 defined in a closed set C_1. By convention, *open sets* are shown as a region delimited by a dashed line and *closed sets* as a region delimited by a solid line.

Definition 1.35 (Topological space) *A set* X *with a topology defined on it, e.g., the pair* $(X, \mathcal{T})$ *is called a topological space. The elements of a topological space are called points.*

Definition 1.36 (Usual topology on $\mathbb{R}$**)** *Given* $X = \mathbb{R}$ *then the usual topology on* $\mathbb{R}$ *is called topology* T *and is defined by:*

$$O \in T \quad iff \quad \forall x \in O \quad \exists \epsilon > 0 :]x - \epsilon, x + \epsilon[\subset O$$

One of the fundamental concepts in a topological space is the concept of neighborhood. Intuitively, the neighborhood of a point is a set which contains the point and other points arbitrarily defined as similar or close. The notion of similarity or closeness can be defined in terms of a metric or rule.

Definition 1.37 (Neighborhood) *A set* U *is called a neighborhood of a point* x *of the topological space* $(X, \mathcal{T})$, *if an open set* $V \in \mathcal{T}$ *exists such that* $x \in V$ *and* $V \subset U$. *The set of neighborhoods of* x *is denoted by* $\mathcal{V}(x)$.

Definition 1.38 (Contact points) *A point* $x \in (X, \mathcal{T})$ *is called a* **contact point** *(also known as closure point or point of closure or adherent point) of a set* $A \in X$, *if every neighborhood of* x *contains at least one point in* A.

Intuitively, given an open set A defined as a region without boundary then the contact points of A are the points of A including its boundary. In Fig. 1.1iv, the point x is a contact point of $A \in X$; thus it is possible to find at least one point p in the neighborhood of x such that $p \in A$.

Definition 1.39 (Limit point) *A point $x \in (X, \mathcal{T})$ is called a **limit point** (accumulation point) of a set $A \subset (X, T)$, if every neighborhood of x contains at least one point of A different from x. In other words, a neighborhood is defined around point $x \in X$ exclusively of elements of A.*

Observe that x does not have to be an element of A; for all neighborhoods $\mathcal{V}$ of x:

$$\mathcal{V} \cap (A \setminus \{x\}) \neq \emptyset$$

In Fig. 1.1iv, the point $y \in X$ but $y \notin A$ (a sort of pinhole on set A) thus y satisfies the properties of a **limit point** in $A \in X$. It is possible to find at least one neighborhood of y exclusively in A. Also y satisfies the axiom of a contact point for A. In general, limit points are contact points but not vice-versa. It can be verified that x is not a limit point: in the neighborhood of x (dotted line) points can always be collected outside A.

Definition 1.40 (Closure) *The set of all contact points of $(X, \mathcal{T})$ is called the **closure** of X and is denoted by $\bar{X}$.*

Definition 1.41 (Interior) *The interior of a non-empty subset $A \subset X$ is defined as the union of all open sets contained in A and is the largest open set contained in it. The interior of A is denoted by **int** A or $\mathring{A}$.*

Definition 1.42 (Weaker / stronger topology) *Let $\mathcal{T}_1$ and $\mathcal{T}_2$ be two topologies defined on the same set X. Topology $\mathcal{T}_1$ is called stronger than topology $\mathcal{T}_2$ (or $\mathcal{T}_2$ is weaker than $\mathcal{T}_1$) if $\mathcal{T}_2 \subset \mathcal{T}_1$. Let τ be the set of all topologies in X. Then for all $T \in \tau$:*

$$\mathcal{T}_t \subset \mathcal{T} \subset \mathcal{T}_d$$

where $\mathcal{T}_t$ is the trivial topology in X and $\mathcal{T}_d$ is the discrete topology in X. $\mathcal{T}_d$ is the strongest topology in X and $\mathcal{T}_t$ is the weakest topology in X.

Theorem 1.1 *Let $\{\mathcal{T}_\alpha\}$ be any set of topologies in X. The intersection $T = \bigcap_\alpha \mathcal{T}_\alpha$ induces a topology in X.*

Definition 1.43 (Trace) *Given a topological space $(X, \mathcal{T})$ and a subset S of X, then the system $\mathcal{T}_S$ defined by:*

$$\mathcal{T}_S = \{S \cap U \mid U \in \mathcal{T}\}$$

*is called the **trace** of S on the set U.*

Definition 1.44 (Density) *Given two sets U and V that are subsets of a topological space $(X, \mathcal{T})$ with $U \subset V$, then the set U is dense in V if $\overline{U} \supset V$, $\overline{U} \in (X, \mathcal{T})$:*

$$x \in V, V \in \mathcal{T} \text{ and open neighborhood of } x, \quad V \cap U \neq \emptyset$$

Let U be a subset of a topological space $(X, \mathcal{T})$. The set U is dense in X if $\overline{U} = X$.

Example: The number π is a real number ($\pi \in \mathbb{R}$) with infinite decimal positions. This number cannot be written as the quotient of two integer numbers ($\pi \notin \mathbb{Q}$). However, it is possible to find an open neighborhood in $\mathbb{Q}$ as small as desired that contains π, for example:

$$\frac{314}{100} < \pi < \frac{315}{100}$$

or in general $\quad \epsilon_1 < \pi < \epsilon_2, \quad \epsilon_1, \epsilon_2 \in \mathbb{Q}, (\epsilon_2 - \epsilon_1) > 0$

Therefore, the rational numbers $\mathbb{Q}$ are dense in $\mathbb{R}$.

1.2.1 Connected topological spaces

One way to differentiate topological spaces is to look at the way they split up into pieces (connectedness). Some basic definitions about connectedness in topological spaces are provided in order to contextualize the space of simplexes which is introduced in section 1.5.1.

Definition 1.45 (Connected set) *A connected set is a set that cannot be partitioned into two non-empty subsets which are open in the relative topology induced on the set. This is a set that cannot be partitioned into two non-empty subsets; each subset has no points in common with the closure of the other.*

Let X be a topological space. The space X is a connected topological space if it is a connected subset of itself [51].

Definition 1.46 (Connected topological space) *A topological space $(X, \mathcal{T})$ is called connected if it is not the union of a pair of disjoint non-empty open sets. The set X is called a connected set [12].*

A connected topological space $(X, \mathcal{T})$ is called unicoherent if the intersection $A \cap B$ is connected for any closed connected sets A, B with $A \cup B = X$.

Definition 1.47 (Connected components) *As a consequence of def. 1.45, the maximal connected subsets (ordered by inclusion) of a non-empty topological space are called connected components of the space.*

The connected components of any topological space X form a partition of X: they are disjoint, non-empty, and their union is the whole space. Every component is a closed subset of the original space. In the case where their number is finite then each component is also an open subset. However, if their number is infinite then this might not be the case; for instance, the connected components of the set of rational numbers are one-point sets which are not open [40].

1.2.2 Continuity and continuous maps

Let f be a mapping from a topological space $(X, \mathscr{T})$ to a topological space $(Y, \mathscr{T})$ meaning that f associates the point $y = f(x) \in Y$ with a point $x \in X$.

Definition 1.48 (Continuity at a point) *The mapping f is said to be continuous at a point $x_0 \in X$, if given any neighborhood $\mathcal{V}_{y_0}$ of the point $y_0 = f(x_0)$ there is a neighborhood $\mathcal{V}_{x_0}$ of the point x_0 such that $f(\mathcal{V}_{x_0}) \subset \mathcal{V}_{y_0}$.*

Definition 1.49 (Continuity on X) *The mapping f is said to be continuous on X if it is continuous at every point of X.*

Theorem 1.2 *The mapping f is continuous iff the inverse image $f^{-1}(\mathcal{V}) \in \mathscr{T}$ of every open set $\mathcal{V}_Y \in \mathscr{T}$ is open, therefore $f^{-1}(\mathcal{V}_Y) \in \mathscr{T}$.*

Theorem 1.3 *The mapping f is continuous iff topology $\mathscr{T}$ is stronger than topology $f^{-1}(\mathscr{T})$.*

Theorem 1.4 *The mapping f is continuous iff the inverse image $f^{-1}(\mathcal{W}_Y)$ of every closed set $\mathcal{W}_Y \subset Y$ is closed in X.*

Fig. 1.1vi shows an example of a continuous mapping. For an open set U_W in $(Y, \mathscr{T}_y)$, any point $x \in f^{-1}(W)$ is associated with $y = f(x)$. Here, $\mathcal{V}_x$ is a neighborhood of x such that $f(\mathcal{V}_x) \subset U_W$.

Theorem 1.5 *Given topological spaces $(X, \mathscr{T})$, $(Y, \mathscr{T})$, and $(Z, \mathscr{T})$, and assuming f to be a continuous mapping of $(X, \mathscr{T})$ to $(Y, \mathscr{T})$ and* *g a continuous mapping of* $(Y, \mathscr{T})$ *to* $(Z, \mathscr{T})$, *then the mapping "∘" is continuous:*

$$g \circ f : x \in X \to g(f(x)) \in Z$$

Definition 1.50 (Homeomorphism) *Given two topological spaces* $(X, \mathscr{T})$ *and* $(Y, \mathscr{T})$, *and letting f be a bijection of* $(X, \mathscr{T})$ *to* $(Y, \mathscr{T})$, *if f and* f^{-1} *are continuous then f is called a* **homeomorphism** *between X and Y. Two topological spaces* $(X, \mathscr{T})$ *and* $(Y, \mathscr{T})$ *are said to be homeomorphic if a homeomorphism exists between them:*

$$(X, \mathscr{T}) \approx (Y, \mathscr{T})$$

An essential property of homeomorphic spaces is that they have the same topological properties, and from a topological point of view they are merely different representations of a single underlying space. If f is a homeomorphism between $(X, \mathscr{T})$ and $(Y, \mathscr{T})$ then $\mathscr{T} = f^{-1}(\mathscr{T})$ and $\mathscr{T} = f(\mathscr{T})$.

Definition 1.51 (Cover) *A* **cover** *of a set A in a topological space* $(X, \mathscr{T})$ *is a family of sets* $\{\mathcal{U}_\alpha\}$ *such that* $A \subset \bigcup_\alpha \mathcal{U}_\alpha$.

Definition 1.52 (Open cover) *A cover* $\{\mathcal{U}_\alpha\}$ *of A in a topological space* $(X, \mathscr{T})$ *is called open if all* $\mathcal{U}_\alpha$ *are open in* $(X, \mathscr{T})$.

Definition 1.53 (Subcover) *A family of sets* $\{V_\alpha\}$ *is called* **subcover** *of A in* $(X, \mathscr{T})$ *if:*

$\{V_\alpha\}$ *is a subset of the cover* $\{\mathcal{U}_\alpha\}$ *of A*

$\{V_\alpha\}$ *is a cover of S*

Definition 1.54 (Compactness) *A topological space* $(X, \mathscr{T})$ *is* **compact** *if every open cover of* $(X, \mathscr{T})$ *has a finite subcover.*

Compactness: The capability of obtaining a cover of set X by a finite number of subsets $\mathcal{U}$ (family of open subsets) allows to represent a compact set by a finite number of set points $\sigma \in \mathcal{U}$. This plays a significant role in computing topological properties for applications.

1.3   METRIC SPACES

At this point metric spaces are introduced in order to pave the way for defining some special topological spaces called *simplicial complexes* as *metric topological spaces*. Metric spaces were first observed by M. Fréchet who studied the properties of a set X equipped with a distance function $d : X \times X \to \mathbb{R}$ called *metric*. The metric notion gives access to the concept of nearness of points and sets.

Definition 1.55 (Distance function) *Let E be a set and $d : E \times E \to \mathbb{R}$ be a function. The distance d on E satisfies the following axioms:*

1. $\forall (x,y) \in E \times E, d(x,y) \geq 0$

2. $\forall (x,y) \in E \times E, d(x,y) = 0$ *iff* $x = y$

3. $\forall (x,y) \in E \times E, d(x,y) = d(y,x)$

4. $\forall (x,y,z) \in E \times E \times E, d(x,y) \leq d(x,z) + d(z,y)$ *(triangular inequality)*

Definition 1.56 (Metric space) *A metric space is a pair (E,d) where E is a set and d is a distance function defined on it.*

Definition 1.57 (Pseudodistance function) *A distance function is called pseudodistance if it does not hold axiom 2 of the distance function. A pseudodistance function spans a space called pseudometric space.*

Definition 1.58 (Quasidistance function) *A distance function is called quasidistance if it does not hold axiom 3 of the distance function. A quasidistance function spans a space called quasimetric space.*

1.3.1   Topology in a metric space

In this section, the reader will understand how topological spaces arise naturally from metric spaces.

Definition 1.59 (Open balls) *Let (E,d) be a metric space and given x in E, then an open ball $B_r(x)$ around x with radius $\epsilon > 0$ is defined by:*

$$B_\epsilon(x) = \{y \in E \mid d(x,y) < \epsilon\}$$

A topology of an open set $\mathcal{O}$ on E is defined by:

$$T = \{\mathcal{O} \in E \mid \forall x \in \mathcal{O} \exists \epsilon > 0, B_\epsilon(x) \subset \mathcal{O}\}$$

The axioms of topology introduced in def. 1.32 can be verified intuitively:

1. *For $\emptyset$ and $X \in T$, if $X = \emptyset$ then any $\epsilon > 0$ is in $\emptyset$. Indeed, $X \in T$ due to the definition of T*

2. *Any arbitrary union of elements of T is in T because the elements of the union of T are such that they already belong to an open ball that can be spanned*

3. *The intersection of a finite number of open sets are points such that a radius r can be selected as small as required*

Theorem 1.6 *In a metric space (X, d) the following axioms are satisfied:*

1. *X and empty set $\emptyset$ are open*

2. *The union of open sets is open*

3. *The finite intersection of open sets is open*

Proof:

1. *Given a small or large ϵ then every point x in X is the center in an open ball $B_\epsilon(x)$. Hence, ϵ is chosen such that $B_\epsilon(x) = X$ is open. Similarly, since there are no points in $\emptyset$, any member that is not in the vicinity of $\emptyset$ cannot be found; hence $\emptyset$ is an open set*

2. *Let $\wedge$ be an index set and suppose that $\{\mathcal{U}_\alpha \mid \alpha \, in \wedge\}$ is a family of open subsets in X and a point x belongs to their union $x \in \{\bigcup_{\alpha \in \wedge} \mathcal{U}_\alpha\}$. Then x must be in $\mathcal{U}_\lambda$ for some $\lambda \in \wedge$ with $\mathcal{U}_\lambda$ being open. There is an open ball $B_\epsilon(x) \subset \mathcal{U}_\lambda$ that is contained in the union $\{\bigcup_{\alpha \in \wedge} \mathcal{U}_\alpha\}$; hence the union is open*

3. *Suppose that $\{\mathcal{U}_n \mid 1 \leq n \leq m\}$ is a finite family of open subsets in X and x belongs to their intersection $x \in \{\bigcap_{1 \leq n \leq m} \mathcal{U}_n\}$. For each n, there is an $\epsilon > 0$ and the open ball contained in $\mathcal{U}_n$. If $\epsilon = \{\mathcal{U}_n \mid 1 \leq n \leq m\}$ then an open ball $\mathcal{U}_n$ is contained in the intersection $\bigcap_{1 \leq n \leq m}$; hence $\bigcap_{1 \leq n \leq m}$ is open*

A strong implication of metric spaces is that - as long as one is able to define a distance function over a set - a topology can be induced on them. However, from a topological space it is not always possible to obtain a metric space.

1.4 ELEMENTS OF AFFINE VECTOR SPACES

The concept of an affine space is described first before going into the theory of geometric simplicial complexes (section 1.5.1).

1.4.1 Affine vector space $\mathcal{A}$

Intuitively and in the words of the mathematical physicist John Baez: *an affine vector space is what is left of a vector space after you have forgotten which point is the origin.* In Fig. 1.2i, the plane P_2 is a kind of 2-dimensional vector space that exhibits a linear structure with no discernible origin; this is an example of an affine vector space.

Definition 1.60 (Vector space) *Given a field $\mathcal{F}$ (def. 1.1.3) then an ordered tuple $(V, +, M)$ where V is a set and M a homomorphism (def. 1.25) is called a vector space over $\mathcal{F}$ if:*

    a. $(V, +)$ is an Abelian group (def. 1.12)

    b. $M : \mathcal{F} \rightarrow Hom(V)$ meaning that M is a function from $\mathcal{F} \times V$ to V as scalar multiplication[2]

An element of V is a vector and elements of $\mathcal{F}$ are scalars.

Definition 1.61 (Affine combination) *Let $S = \{p_0, p_1, ..., p_k\}$ be a subset of a field $\mathcal{F}$. A linear combination of S and coefficients λ_i is given by:*

$$x = \sum_{i=0}^{k} \lambda_i p_i \qquad \lambda_i \in \mathbb{R}$$

An affine combination is a linear combination in which the sum of the coefficients is 1:

$$\sum_{i=0}^{k} \lambda_i = 1$$

---

[2]Scalar multiplication refers to the multiplication of a vector by a constant s producing a vector in the same ($s > 0$) or opposite ($s < 0$) direction but of different length (Weisstein, Eric W. "Scalar Multiplication." From MathWorld–A Wolfram Web Resource. http://mathworld.wolfram.com/ScalarMultiplication.html)

An affine combination induces a linear subspace called an *affine space* which is nothing more than a vector space under affine combinations. The vector space and the affine automorphisms (linear combinations from a set to itself) induce a group called the *affine group*.

Definition 1.62 (Affine space $\mathcal{A}$) *An affine vector space $\mathcal{A}$ over a field F is a tuple (A, V, ϕ) where A is an non-empty set, V is a vector space, and $\phi : A \times A \longrightarrow V$ is an application that satisfies:*

    a. $\forall P \in A$ *and* $\forall \vec{u} \in V$ *exists a unique* $Q \in A$ *such that*
        $\phi(P, Q) = \vec{u}$

    b. $\phi(P, Q) + \phi(Q, R) = \phi(P, R)$ *for all* $P, Q, R \in A$

The elements of A are called points and V is a vector space associated to the affine space (A, V, ϕ). The dimension in the affine vector space (A, V, ϕ) is defined by:

$$dim A = dim V$$

Properties: An affine space $\mathcal{A}$ satisfies the following:

    1. $\phi(P, Q) = \vec{0}$ *iff* $P = Q$

    2. $\phi(P, Q) = -\phi(Q, P), \forall P, Q \in \mathbb{A}$

    3. $\phi(P, Q) = \phi(R, S)$ *iff* $\phi((P, S) = \phi(R, Q)$

Every ordered pair of points P and Q in an affine space is associated with a vector $\overrightarrow{PQ}$ by applying $\phi(P, Q) = \overrightarrow{PQ} = Q - P$. In the affine space, vectors and points are interchangeable terms.

Definition 1.63 (Origin $\mathcal{O}$) *Given an affine space of dimension n, namely $\mathcal{A} = (A, V, \phi)$ and a set of $n + 1$ points $P = \{P_0, P_1, \ldots, P_n\}$ such that $P \in A$, then P is an affine system of reference for $\mathcal{A}$ if the set of vectors $\{\overrightarrow{P_0 P_1}, \ldots, \overrightarrow{P_0 P_n}\}$ is a base for vector space V. The point $P_0 \in A$ such that $\{\overrightarrow{P_0 P_1}, \ldots, \overrightarrow{P_0 P_n}\}$ is a base for vector space V called the origin of the reference system $\{P_0; P_1, \ldots, P_n\}$. The origin point is denoted by $\mathcal{O}$. An origin point $\mathcal{O} \in A$ and a base for vector space $B \in V$ build together an affine reference system or affine coordinate system denoted by $\mathcal{R} = \{\mathcal{O}; B\}$.*

Definition 1.64 (Affinely independent points) *Given an affine space of dimension n, namely $\mathcal{A} = (A, V, \phi)$ and a set of $n + 1$ points $P = \{p_0, p_1, \ldots, p_n\}$, if the unique solution for the linear combination:*

$$\sum_{i=0}^{k} \lambda_i p_i = 0 \qquad \lambda_i \in \mathbb{R}$$

*is $\lambda_i = 0 \quad \forall i \in \{1, 2, \ldots, k\}$ then the points in P are **linearly independent**. If the unique solution for the affine combination:*

$$\sum_{i=0}^{k} \lambda_i p_i = 0 \qquad \sum_{i=0}^{k} \lambda_i = 1 \qquad \lambda_i \in \mathbb{R}$$

*is $\lambda_i = 0 \quad \forall i \in \{1, 2, \ldots, k\}$ then the points in P are **affinely independent**. Linear independence implies affine independence but not vice versa.*

Definition 1.65 (Convex combination) *Let $S = \{p_0, p_1, \ldots, p_k\}$ be a set of points in a field $\mathcal{F}$. An affine combination of S and λ_i coefficients in which all of the coefficients are strictly non-negative is called a convex combination:*

$$x = \sum_{i=0}^{k} \lambda_i p_i, \qquad \lambda_i \geq 0 \quad and \quad \sum_{i=0}^{k} \lambda_i = 1$$

Definition 1.66 (Convex set) *Given an affine space $\mathcal{A} = (A, V, \phi)$ then a set of points $C \in \mathcal{A}$ is called a convex set iff for any $p_1, p_2 \in C$ and $0 \leq \lambda \leq 1$:*

$$\lambda p_1 + (1 - \lambda) p_2 \in C$$

This linear sum is also called a convex combination of p_1 and p_2.

Fig. 1.1vii presents examples of a) a convex set where all convex combinations are in set C_1, and b) a non-convex set where some convex combinations lie outside set C_2.

Definition 1.67 (Convex hull) *The convex hull of a finite set of points C in an affine space $\mathcal{A}$ with all possible convex combinations of subsets of C is denoted by:*

$$conv(C) = \left\{ \sum_{j=1}^{N} \lambda_j p_j : \lambda_j \geq 0 \, \forall j \, and \, \sum_{j=1}^{N} \lambda_j = 1 \right\}$$

The convex hull $conv(C)$ is the smallest convex set containing C.

In Fig. 1.1viii, the points $\{p_1, p_2, p_3, p_9, p_{10}\}$ define the convex hull of a set of points.

1.5 SIMPLICIAL COMPLEXES

The aim of the section is to introduce an important class of topological spaces: simplicial complexes. These are useful spaces since they allow the subdivision of a complicated topological space into simple building elements that when pieced together construct a sort of discrete approximation of the original space. Such an approximation is formed by a discrete and denumerable set of elements called simplexes or simplices. Simplexes can be efficiently handled computationally thus allowing the computation of topological properties.

In this monograph, simplicial complexes are used as a discretized representation of the topological space of images for obtaining topological features. Simplicial complexes are described from a geometric point of view, but at the end of the chapter their pure combinatorial structure is provided as well.

1.5.1 Geometric simplicial complexes

Geometric simplicial complexes are topological spaces comprised of finite sets of points, line segments, triangles, and the triangle generalization into higher dimensions. These constituents are called simplexes. The simplexes of geometric simplicial complexes are embedded in an affine vector space.

Definition 1.68 (Simplex) *A geometrical simplex is a set of ordered points in an affine space $\mathcal{A}$. Often points are referred to as vertices. A simplex is said to be proper when its vertices are linearly independent and improper when there is a linear dependence between its points.*

Definition 1.69 (k-simplex) *A k-simplex σ is a set of points defined by the convex hull spanned by $k + 1$ geometrically affine independent points:*

$$\sigma = conv(\{v_0, v_1, \ldots, v_k\}) = \left\{ \sum_{j=0}^{N} \lambda_j v_j : \lambda_j \in \mathbb{R}, \lambda_j \geq 0 \forall j, \sum_{j=0}^{N} \lambda_j = 1 \right\}$$

The points v_j are often called vertices and the values λ_j are called the barycentric coordinates of point v_j with respect to σ. Here, $\{v_i\}$ spans σ and its dimension is $dim\ \sigma = k$. Recall that affine independent points refer to the collection of n vectors $\{(v_1 - v_0), \ldots, (v_n - v_0)\}$ being linearly independent.

Fig. 1.2iii presents an example of k-simplexes generated by the convex hull of two sets of vertices: a) the 1-simplex $\sigma = \{v_1, v_2\}$ spans a line segment between v_1 and v_2 when coefficient λ varies between $[0,1]$, and b) the 2-simplex $\sigma = \{v_1, v_2, v_3\}$ spans a filled triangle.

Specific names are given for the first simplex dimensions: vertex for the *0-simplex*, edge for the *1-simplex*, triangle for the *2-simplex*, and tetrahedron for the *3-simplex* (Fig. 1.2iv). A nice property of simplicial complexes is that a *k-simplex* is homeomorphic to a closed *k-dimensional* ball, meaning that a function exists that can continuously shape a *0-simplex* into a *0-ball* (point), a *1-simplex* into a *1-ball* (line segment), a *2-simplex* into a *2-ball* (disk), and a *3-simplex* into a *3-ball* (ball; interior of a sphere). Simplexes are useful building blocks for building up topological spaces called *simplicial complexes*. Simplicial complexes are the framework where homology is deployed to measure their properties.

Definition 1.70 (Face τ of a k-simplex) *Any subset of an affinely independent set of points is affinely independent and can span a new simplex. A face of a simplex σ is another simplex σ' whose vertices are also vertices of σ and therefore $\sigma' \in \sigma$. A face of a k-simplex σ is the convex hull of an inhabited subset of points of σ denoted as τ. When τ is a proper subset of σ then it spans a **proper face** ($\tau < \sigma$) otherwise it is just a face ($\tau \leq \sigma$). Since a set of size $(k + 1)$ has 2^{k+1} subsets including the empty set, a k-simplex σ with $k + 1$ elements spans exactly $(2^{k+1}-1)$ proper faces; this is the power set $\wp$ of its vertices $\wp\{v_0, v_1, \ldots, v_k\}$ without σ itself (Fig. 1.3i).*

Definition 1.71 (Coface) *Given a simplex σ and a subset τ ($\tau \in \sigma$) which is a face of σ, then σ is a coface of τ. If $\sigma \in K$ and the only coface of σ is itself, then σ is called a maximal simplex.*

Geometric simplicial complexes

The idea of simplicial complexes is to build up a space by the union of simplexes that have different dimensions.

Definition 1.72 (Geometric simplicial complex) *A simplicial complex K is a finite collection of k-simplexes in a geometric affine space $\mathcal{A}$ which satisfies:*

1. For any simplex $\sigma \in K$ every face of σ is a face of K

2. *Given two simplexes* $\sigma, \tau \in K$:

$$\sigma \cap \tau = \begin{cases} \emptyset \\ \sigma' & ,if\ \sigma' \in K \end{cases}$$

The underlying space of a simplicial complex K is denoted by $|K|$ and it is the topological space of the union of its simplexes:

$$|K| = \bigcup_{\sigma \in K} \sigma$$

Each simplex is considered as a topological subspace. A triangulation of a topological space $(X, \mathscr{T})$ is a simplicial complex K such that $|K| \approx (X, \mathscr{T})$. The underlying space of K is homeomorphic to $(X, \mathscr{T})$ which is triangulable when K exists. Triangulations can represent topological spaces as simplicial complexes. Finally, two complexes K and L are isomorphic if $|K| \approx |L|$. Fig. 1.2ii presents a simplicial complex; simplexes of different dimensions are glued together to cover a space.

1.5.2 Oriented simplicial complexes

In a geometric simplex, convex combinations of vertices span vectors which are oriented depending on the order of appearance of the vertices. Recall that vertices in an affine space are essentially vectors. Orientation is an inherent property of geometric simplexes. The explicit order of simplexes is denoted by parentheses; for instance consider $\sigma = (v_0, v_1)$ meaning that simplex σ has two faces and the orientation of the simplex is such that it starts at v_0 and it ends at v_1. Only two ways exist to orientate a simplex which means that only two disjoint classes of orientations exist. This is illustrated with an example below.

Example: Given a $2 - simplex\ \sigma = \{v_0, v_1, v_2\}$ (Fig. 1.3iii) then the following orderings are considered:

- $(v_0, v_1, v_2), (v_1, v_2, v_0), (v_2, v_0, v_1)$ have the same orientation

- $(v_1, v_0, v_2), (v_0, v_2, v_1), (v_2, v_1, v_0)$ have the same orientation

- $(v_0, v_1, v_2) \not\equiv (v_1, v_0, v_2)$

It is possible to observe a relationship between the way vertices are permuted and their equivalence.

An even permutation defines an equivalence relation on the set of all possible orderings. Two orderings are equivalent if they differ by an even permutation. An even permutation is a rearrangement of the elements of an ordered set that is obtained by an even number of two element swaps, e.g., for a set of n elements ($n > 2$) there exist $n!/2$ even permutations. The orientation of a simplex is the arbitrary selection of an equivalence class of vertex orderings.

Definition 1.73 (Oriented simplex) *An oriented simplex is a simplex with a choice of orientation. The two possible orientations of a simplex are often designated with the signs "+" and "-".*

One convention for selecting an orientation is to use the determinant of the vertex basis orientation called natural orientation.

Definition 1.74 (Natural orientation) *Given an oriented simplex $\sigma = \{v_0, v_1, \ldots, v_n\}$ by vertex ordering $(v_0, v_1, \ldots, v_n)$, then the natural orientation of the simplex is given by the sign function:*

$$sgn(\sigma) = det(v_1 - v_0, v_2 - v_0, \ldots, v_n - v_0)$$

Given the orientation $sgn(\sigma) = sgn(\{v_1, v_2\})$ then the opposite orientation is $sgn(\{v_2, v_1\})$ or $\{v_1, v_2\} = -\{v_2, v_1\}$.

Building blocks of oriented simplicial complexes

Definition 1.75 (0-simplex) *A $0-simplex$ is a vertex or point (Fig. 1.2va):*

$$0 - simplex = (v_0)$$

Definition 1.76 (1-simplex) *A $1-simplex$ is an oriented edge (Fig. 1.2vb):*

$$1 - simplex = (v_0, v_1)$$

with (v_0, v_1) denoting that the edge point from v_0 to v_1 is a $1 - simplex$.

Definition 1.77 (2-simplex) *A $2 - simplex$ is a two-dimensional oriented face (Fig. 1.2vc):*

$$
\begin{aligned}
2\text{-}simplex\ \sigma &= (v_0, v_1, v_2) = (v_1, v_2, v_0) = (v_2, v_0, v_1) \\
-\sigma &= (v_1, v_0, v_2) = (v_0, v_2, v_1) = (v_2, v_1, v_0)
\end{aligned}
$$

Definition 1.78 (3-simplex) *A 3 − simplex is an oriented solid tetrahedral (Fig. 1.2vd) that induces 12 equivalent orientations:*

$$
\begin{aligned}
3\text{-}simplex \; \sigma \;&= (v_0, v_1, v_2, v_3) = (v_1, v_2, v_0, v_3) = (v_2, v_0, v_1, v_3) \\
&= (v_1, v_0, v_3, v_2) = (v_2, v_1, v_3, v_0) = (v_0, v_2, v_3, v_1) \\
&= (v_3, v_1, v_0, v_2) = (v_3, v_2, v_1, v_0) = (v_3, v_0, v_2, v_1) \\
&= (v_0, v_3, v_1, v_2) = (v_1, v_3, v_2, v_1) = (v_2, v_3, v_0, v_1) \\
-\sigma \;&= (v_1, v_0, v_2, v_3) = (v_2, v_1, v_0, v_3) = (v_0, v_2, v_1, v_3) \\
&= (v_1, v_3, v_0, v_2) = (v_2, v_3, v_1, v_0) = (v_0, v_3, v_2, v_1) \\
&= (v_0, v_1, v_3, v_2) = (v_1, v_2, v_3, v_0) = (v_2, v_0, v_3, v_1) \\
&= (v_3, v_0, v_1, v_2) = (v_3, v_1, v_2, v_0) = (v_3, v_2, v_0, v_1)
\end{aligned}
$$

Summary

A set of convenient building blocks called *k-simplexes* - when put together by the union of sets - build a topological space called a *simplicial complex*. This is a geometric approach meaning that the simplexes are embedded in an affine vector space, e.g., a Euclidean space. Simplexes have the property of orientation implying that their order is important. A simplex spanned by two vertices has the following notation regarding its orientation: $(v_1, v_2) = -(v_2, v_1)$.

Later in the chapter, elements of group theory are introduced for organizing simplexes, simplicial complexes, and their orientations in a system of simple algebraic patterns (repeating sequences of operations and symbols) and symmetries (mappings of an object onto itself preserving the structure).

1.5.3 Unoriented simplicial complexes

An unoriented simplex is a simplification of the geometrical simplicial complex concept. Orientation is sacrificed and only the sense of vicinity remains. A simplex σ is considered equivalent to its oppositely oriented $-\sigma$. Intrinsically, this means that coefficients 1 and −1 associated with orientation have to be equivalent $1 = -1$. So rather than being coefficients in $\mathbb{Z}^n$ or $\mathbb{R}^n$ they are situated in $\mathbb{Z}_2$. A satisfactory definition for simplicial complexes can be developed based on $\mathbb{Z}_2$ thanks to homology theory introduced by H. Tietze [45].

The motivation of including this definition is that it is interesting for classifying topological spaces according to their homology. As in the case of graphs, the particular orientation given to a simplicial complex does not affect homology groups up to isomorphism [19]. Although it is a simplified approach that sacrifices details about space, simplicial complexes with $\mathbb{Z}_2$ coefficients are easier to compute. To start let's recall that $\mathbb{Z}_2$ (group of integers modulo 2) is spanned by a set of two elements: integer numbers $\{0,1\}$ and addition as the group operator. A free vector space over the field $\mathbb{Z}_2$ generated by a set $S = \{x_1, x_2, \ldots, x_k\}$ consists of all elements of the form:

$$n_1 x_1 + n_2 x_2 + \ldots + n_k x_k$$

where n_i in $\mathbb{Z}_2$ (see example 1.1.3).

Building blocks of unoriented simplicial complexes

The first four dimensions of unoriented simplicial complexes are presented (Fig. 1.2iv).

Definition 1.79 (0-simplex) *The 0-simplex is a vertex (Fig. 1.2iva) and because orientation is not present, simplexes are represented as sets:*

$$0 - simplex = point = \{v_0\}$$

Definition 1.80 (1-simplex) *The 1-simplex is an edge (Fig. 1.2ivb):*

$$1 - simplex = edge = \{v_0, v_1\}$$

The 1-simplex is composed of two 0-simplexes: $\{v_0\}$, $\{v_1\}$.

Definition 1.81 (2-simplex) *The 2-simplex is a solid triangle (Fig. 1.2ivc):*

$$2 - simplex = face = \{v_0, v_1, v_2\}$$

The 2-simplex is composed of three 1-simplexes: $\{v_0, v_1\}$, $\{v_2, v_3\}$, $\{v_1, v_3\}$.

Definition 1.82 (3-simplex) *The 3-simplex is a solid tetrahedral (Fig. 1.2ivd):*

$$3 - simplex = tetrahedron = \{v_0, v_1, v_2, v_3, v_4\}$$

The 3-simplex is composed of four 2-simplexes: $\{v_0, v_1, v_3\}$, $\{v_1, v_2, v_3\}$, $\{v_0, v_1, v_2\}$, $\{v_0, v_2, v_3\}$.

This paradigm of defining simplicial complexes continues to higher dimensions.

1.5.4 Abstract simplicial complexes

Up to this point, simplicial complexes were presented as embedded in an affine space that defines their geometry. Sacrificing the orientation of simplexes by laying complexes on $\mathbb{Z}_2$ is beneficial for concentrating on more fundamental properties of the space: *its connectivity.*

In this section, the concepts of affine space and geometry of simplexes are dropped in order to define a simplicial complex as a system of sets of vertices. The aim of this approach is to work on the fundamental combinatorial structure of simplexes and to obtain a generalization for simplicial complexes without geometry. This does not mean that the objects have no geometric meaning but there is no defined geometry.

Definition 1.83 (Abstract simplicial complex) *An abstract simplicial complex is a pair $K = (V_K, S_K)$ given by a finite set of points V_k called vertices of K, and S_K which is a set of non-empty finite subsets of V_k called simplexes of K such that:*

a. $(\forall v \in V_k), \{v\} \in S_k$

b. $(\forall \sigma \in S_k), (\forall \sigma' \subset \sigma, \sigma' \neq \emptyset), \sigma' \in S_k$

Example: *Consider $V_k = \{a, b, c, d, e\}$ then a valid set of simplexes of V_k is:*

$$S_k = \{\{a\}, \{b\}, \{c\}, \{d\}, \{e\}, \{a,b\}, \{a,c\}, \{a,d\}, \{b,c\}, \{b,d\},$$
$$\{c,d\}, \{c,e\}, \{d,e\}, \{a,b,c\}, \{a,b,d\}, \{a,c,d\}, \{b,c,d\},$$
$$\{c,d,e\}, \{a,b,c,d\}\}$$

$K = (V_K, S_K)$ is an abstract simplicial complex spanned by V_K and S_K.

Definition 1.84 (p-simplex) *A simplex σ of cardinality $p+1$ $p \geq 0$ having $(p + 1)$ elements is called a p-simplex of dimension p:*

$$dim \sigma = p$$

As a consequence of this definition a 0-simplex is a vertex.

Definition 1.85 (Face of an abstract simplicial complex) *If σ is a simplex of K, every non-empty $\sigma' \subset \sigma$ is a face of σ. A p-simplex σ spans exactly $(2^{p+1}-1)$ proper faces which is the power set of its vertices $\mathcal{P}(\sigma)$ without σ. All faces of a simplex are also simplexes.*

Definition 1.86 (Dimension of an abstract simplicial complex)
The dimension of an abstract simplicial complex K is defined by:

$$dim\,K = \max\{dim\,\sigma \mid \sigma \in K\}$$

Definition 1.87 (Subcomplex) *Given two simplicial complexes $K = (V_k, S_k)$ and $L = (V_l, S_l)$, then L is a subcomplex of K if $V_l \subset V_k$ and $S_l \subset S_k$.*

Definition 1.88 (Generated complex) *Let $K = (V_k, S_k)$ be a simplicial complex then for every simplex $\sigma \in S_k$:*

$$\bar{\sigma} = (\sigma, \mathcal{P}(\sigma) \setminus \emptyset)$$

where $\mathcal{P}(\sigma)$ is the power set of vertices in σ. The simplicial complex $\bar{\sigma}$ is generated by σ.

Definition 1.89 (Join of abstract simplicial complexes) *Given two simplexes τ and σ of K then the join of τ and σ is denoted by $\tau * \sigma$ and is defined by $\tau * \sigma = \tau \cup \sigma$ which is the union of their vertices.*

1.6 HOMOLOGY

The approximation of a topological space by gluing simple pieces together is an effective way of obtaining the intrinsic properties of the space. In the field of topology, these properties are called topological invariants. A topological invariant is a feature of the space that is preserved under the action of a continuous function over it. Here, an important invariant is introduced: *the homology group*.

SIMPLICIAL HOMOLOGY

Homology is an analytical tool that quantifies certain algebraic invariants like the homology groups in a topological space. Homology groups characterize the number and type of holes in a topological space, thus providing a fundamental description of its structure. This information can be used, e.g., to describe an image or determine the similarities between images. It is appropriate to mention that homology groups do not capture all topological aspects of a space since two spaces with the same homology groups may not be topologically equivalent. However, two spaces that are topologically equivalent must have equal homology groups.

When the subject of analysis is a point cloud (set of data points in space, e.g., the pixels of an image), a conventional procedure to harvest homology groups consists of generating an approximation of the subjacent topology of the data. Such an approximation is often based on simplicial complexes. The problem of computing homology groups from regular cell complexes such as simplicial ones is well-known and solved in the literature. The challenge though is to obtain the approximation in a way that captures all the interesting aspects of the space. As shown above, approximations can be done in different ways, e.g., using oriented simplicial complexes, unoriented simplicial complexes, or abstract simplicial complexes. In this section, the homology calculations for all of them are presented.

1.6.1 Homology of oriented simplicial complexes

An introduction to oriented simplicial complexes (Fig. 1.2v) is given in section 1.5.2. The homology of a space of simplicial complexes may be regarded as the algebraization of how simplexes of dimension n attach to simplexes of dimension $n - 1$ [21]. This leads to the concepts of chain, chain group, boundary, and boundary group for obtaining the homology groups.

Definition 1.90 (A *p-chain* of an oriented p-simplex) *Let K be an oriented complex. A **p-chain** denoted by c_p of K is defined by the formal linear combination:*

$$c_p = \sum_{i=0}^{p} \alpha_i \sigma_i \qquad (1.2)$$

of oriented p-simplexes σ_i of K added with coefficients α_i of rational integers.

Thus, a chain can be visualized as a collection of oriented simplexes taken with certain multiplicities [22]. In the chain 1.2, the oriented simplex σ_i has multiplicity α_i ($\alpha_i \geq 0$) otherwise the simplex $-\sigma_i$ has multiplicity $-\alpha_i$. An example is shown in Fig. 1.3ii. Conveniently, *p-chains* can be used as generators of free Abelian groups.

Definition 1.91 (The *p*-chain group) *This is the free Abelian group of chains having dimension p in a simplicial complex K under the addition operator +:*

a. *The identity of this group is a chain such that $0 = \sum_{i=1}^{k} 0 * \sigma_i$ with σ_i being a simplex of dimension p*

b. *The inverse of a p-chain is a p-chain in the opposite orientation since $(v_0, v_1, \ldots, v_p) = -(v_1, v_0, \ldots, v_p) \implies (v_0, v_1, \ldots, v_p) + (v_1, v_0, \ldots, v_p) = 0$*

The p-chain group in a simplicial complex K is denoted by C_p or $C_p(K)$.

Definition 1.92 (Boundary of an oriented p-simplex) *The boundary of a simplex is the union of all the **proper faces** of a simplex. Let $\sigma = (v_0, v_1, \ldots, v_{p+1})$ be an oriented (p+1)-simplex spanned by vertices $\{v_0, v_1, \ldots, v_{p+1}\}$. The boundary of σ denoted by $\partial(\sigma)$ is defined by:*

$$\partial_{p+1}(\sigma) = \sum_{i=0}^{p+1} (-1)^i (v_0, \ldots, \widehat{v_i}, \ldots, v_{p+1})$$

This is the alternating sum of oriented p-faces or p-simplexes based on the removal of one vertex at a time sequentially. In Fig. 1.3ii, the elements of C_1 are obtained by integral linear combinations of *1-chain* edges $\{e_1, e_2, e_3\}$. The boundary of these *1-chains* in K is:

$$\partial_1(e_1) = v_2 - v_1; \quad \partial_1(e_2) = v_3 - v_2; \quad \partial_1(e_3) = v_1 - v_3; \quad \partial_1(e_4) = v_1 - v_3$$

The boundary is a group homomorphism that maps one-dimension edges into zero-dimension vertices $\partial_1 : C_1 \to C_0$. Given a general element of C_1: $c = \alpha_1 e_1 + \alpha_2 e_2 + \alpha_3 e_3 + \alpha_4 e_4$ then its boundary, using the properties of homomorphism, is:

$$
\begin{aligned}
\partial(c) \quad &= \alpha_1 \partial(e_1) + \alpha_2 \partial(e_2) + \alpha_3 \partial(e_3) + \alpha_4 \partial(e_4) \\
\bullet \quad &= \alpha_1(v_2 - v_1) + \alpha_2(v_3 - v_2) + \alpha_3(v_1 - v_3) + \alpha_4(v_1 - v_3) \\
\bullet \quad &= (\alpha_3 + \alpha_4 - \alpha_1)v_1 + (\alpha_1 - \alpha_2)v_2 + (\alpha_2 - \alpha_3 - \alpha_4)v_3
\end{aligned}
$$

This is the boundary in terms of the basis (v_1, v_2, v_3).

Definition 1.93 (Boundary map) *Let K be a simplicial complex of dimension ω. For $1 \leq p \leq \omega$, the following mapping is defined:*

$$\partial_p = \partial : C_{p+1}(K) \to C_p(K)$$

with $C_{p+1}(K)$ being the p+1-chain in K and $C_p(K)$ being the p-chain in K. If a formal linear combination of simplexes of dimension p $\sum_{i=0}^{n} \alpha_i \sigma_i$ exists as a p-chain in K, then:

$$\partial(\sum_{i=0}^{n} \alpha_i \sigma_i) = \sum_{i=0}^{n} \alpha_i \partial(\sigma_i)$$

For all other p, ∂_p will be the zero map. The maps ∂_p are called boundary maps. Notice that each ∂_p is a linear map.

Definition 1.94 (p-cycles) *Let K be a simplicial complex in a field $\mathcal{F}$. For $1 \leq p \leq dim K$, the kernel of the boundary map:*

$$\partial_q : C_p \to C_q, \quad q < p \quad, often \ q = p - 1$$

are the elements of C_p such that under ∂ they become the identity of C_q which is 0 in the p-chain group. The elements of the kernel are called p-cycles and are denoted by Z_p.

The concept is illustrated in Fig. 1.3ii. Given the elements of $C_1 \in K$ then a particular linear combination is observed $c = 1 * e_1 + 1 * e_2 + 1 * e_3 + 0 * e_4 = e_1 + e_2 + e_3$. This combination is a closed loop on K with boundary:

$$\begin{aligned} \partial(c) \quad &= \partial(e_1) + \partial(e_2) + \partial(e_3) \\ \bullet \quad &= (v_2 - v_1) + (v_3 - v_2) + (v_1 - v_3) \\ \bullet \quad &= (v_2 - v_2) + (v_1 - v_1) + (v_3 - v_3) = 0 \quad \text{(the group identity)} \end{aligned}$$

This is an algebraic way to identify a cycle. The 1-cycle in the space K of the graph (Fig. 1.3ii) is any linear combination of the 1-chain $c \in C_1$ such that $\partial(c) = 0$. Considering c as a general linear combination of the 1-chain in K then the boundary of c is:

$$\partial(c) = (\alpha_3 + \alpha_4 - \alpha_1)v_1 + (\alpha_1 - \alpha_2)v_2 + (\alpha_2 - \alpha_3 - \alpha_4)v_3 = 0$$

The cycles on K are given by the solutions of the following linear system of equations:

$$\begin{aligned} \alpha_3 + \alpha_4 - \alpha_1 &= 0 \\ \alpha_1 - \alpha_2 &= 0 \\ \alpha_2 - \alpha_3 - \alpha_4 &= 0 \end{aligned} \qquad \text{Solutions} \to \qquad \begin{aligned} &\text{a) } \alpha_1 = \alpha_2 = \alpha_3 = 1 \\ &\text{b) } \alpha_1 = \alpha_2 = \alpha_4 = 1 \end{aligned}$$

The solutions form a basis for the space of cycles. In summary, using the boundary map $\partial : C_1 \rightarrow C_0$ a kernel group $\partial(C_1)$ of elements $\langle (\alpha_1 + \alpha_2 + \alpha_2), (\alpha_1 + \alpha_2 + \alpha_4) \rangle$ is obtained which is *isomorphic to $\mathbb{Z} \times \mathbb{Z}$*:

$$Ker\partial(C_1) \cong \mathbb{Z}^2$$

Two major groups (free Abelian groups) arise from p-cycles:

1. All p-cycles form a group called the **p-cycle group** $Zp := Zp(K)$ under addition as the chain group

2. The image of $\partial_{p+1} : C_{p+1} \rightarrow C_p$ forms a group of **p-boundaries** denoted by $Bp(K) = B_p$, $B_p \subseteq Z_p \subseteq C_p$

Fig. 1.3iv presents a diagrammatic representation of Abelian groups of the chain, cycle, and boundary under mapping ∂.

Definition 1.95 (Homology group) *A consequence of def. 1.94 is that the group B_p is a subgroup of Z_p. The simplicial p-th homology group of K, namely $H_p(K) = H_p$ or homology p-group for short is defined by the following quotient group:*

$$H_p = Z_p / B_p$$

Definition 1.96 (Betti number) *Let K be a simplicial complex then this number:*

$$\beta_p := rank(H_p(K))$$

is called the Betti number named after the mathematician Enrico Betti. Recall that the rank of an Abelian group is the cardinality of the maximal linearly independent subsets of the group's generators.

The invariant Euler characteristic is linked to Betti numbers by the relation:

$$\chi(K) = \sum_{p=0}^{dimK} (-1)^p \beta_p$$

Informally, the p-th Betti number refers to the number of k-dimensional holes on a topological space. The first few Betti numbers have the following definitions for the *0-dimensional, 1-dimensional*, and *2-dimensional* simplicial complexes:

β_0 is the number of connected components

β_1 is the number of one-dimensional or "circular" holes

β_2 is the number of two-dimensional "voids" or "cavities"

1.6.2 Homology of unoriented simplicial complexes

Unoriented simplicial complexes are complexes embedded on $\mathbb{Z}_2$. This is a group spanned by a set of two elements: integer numbers $\{0,1\}$ and addition as the group operator (called *symmetric difference*). The description of this homology group is equivalent to the homology of oriented simplicial complexes but embedded in $\mathbb{Z}_2$.

Definition 1.97 (Symmetric difference) *The addition operator on a vector space over the field $\mathbb{Z}_2$ is called the **symmetric difference** denoted by $\bigoplus$ (Table 1.1).*

The symmetric difference of two sets is the set of elements which are in either of the sets and not in their intersection.

Table 1.1: Cayley table showing the symmetric difference operator.

$\bigoplus$	0	1
0	0	1
1	1	0

Unoriented p-chain

A *p-chain* is a formal linear combination of simplexes σ of dimension p, $c_a = \sum_{i=0}^{p} \alpha_i \sigma_i$, where coefficients lay on $\mathbb{Z}_2$ and the addition operator is $\bigoplus$.

Unoriented p-chain group C_p

The set of unoriented *p-chains* of a simplicial complex K forms the Abelian group C_p.

Boundary operator

To obtain the boundary of an unoriented simplex it is important to note that in $\mathbb{Z}_2$ this relation exists: $-1=1$, therefore the alternating sum of def. 1.92 becomes the sum of simplex faces in $\mathbb{Z}_2$:

$$\partial_p(\sigma) = \sum_{i=0}^{p} (v_0,\ldots,\widehat{v_i},\ldots,v_p)$$

where $\widehat{v_i}$ means that one vertex is removed from the sum at each time. Since addition is commutative it does not matter which vertex is removed first. To illustrate this consider Fig. 1.2iv; $\sigma_{\{x_0, x_2, \dots x_p\}}$ is the *p-simplex* spanned by $\{x_0, x_2, \dots x_p\}$ vertices:

0-simplex: vertex $\sigma\{v_0\}$; the boundary removes one vertex at a time ($\partial \sigma = 0$) because only one element is involved (Fig. 1.2iva)

1-simplex: edge σ spanned by vertices $\{v_0, v_1\}$; the boundary is the sum of its vertices removing one at a time ($\partial \sigma = v_1 \oplus v_0$) (Fig. 1.2ivb)

2-simplex: face σ spanned by vertices $\{v_0, v_1, v_2\}$; the boundary is the sum of its vertices removing one at a time ($\partial \sigma = \sigma_{\{v_0, v_1\}} \oplus \sigma_{\{v_1, v_2\}} \oplus \sigma_{\{v_0, v_2\}}$) (Fig. 1.2ivc)

3-simplex: solid tetrahedron $\sigma = \{v_0, v_1, v_2, v_3\}$; the boundary is the sum of its vertices removing one vertex at a time ($\partial \sigma = \sigma_{\{v_0, v_1, v_2\}} \oplus \sigma_{\{v_0, v_1, v_3\}} \oplus \sigma_{\{v_0, v_2, v_3\}} \oplus \sigma_{\{v_1, v_2, v_3\}}$) (Fig. 1.2ivd)

Unoriented *p*-cycle

By def. 1.91, a *p-cycle* c is a *p-chain* with an empty boundary $\partial_p c = 0$ where 0 is the identity in C_{p-1}. In the next example, this is derived under $\mathbb{Z}_2$ and operator $\oplus$.

Example: Consider the problem of obtaining the boundary of a *2-simplex* σ spanned by vertices $\{v_0, v_1, v_2\}$ (Fig. 1.2ivc):

$$\partial \sigma = \sigma_{\{v_0, v_1\}} \oplus \sigma_{\{v_1, v_2\}} \oplus \sigma_{\{v_0, v_2\}}$$

The boundary of the face is the formal linear combination of simplexes: *1-chain*. The boundary of the *1-chain* is computed as follows:

$$
\begin{aligned}
\partial(\sigma_{\{v_0, v_1\}} \oplus \sigma_{\{v_1, v_2\}} \oplus \sigma_{\{v_0, v_2\}}) &= \partial \sigma_{\{v_0, v_1\}} \oplus \partial \sigma_{\{v_1, v_2\}} \oplus \partial \sigma_{\{v_0, v_2\}} \\
&= v_0 \oplus v_1 \oplus v_1 \oplus v_2 \oplus v_0 \oplus v_2 \\
&= v_0 \oplus v_0 \oplus v_1 \oplus v_1 \oplus v_2 \oplus v_2 \\
&= 0 \quad \text{by the properties} \\
&\qquad \text{of the operator } \oplus
\end{aligned}
$$

Thus, the *1-chain* $\sigma_{\{v_0, v_1\}} \oplus \sigma_{\{v_1, v_2\}} \oplus \sigma_{\{v_0, v_2\}}$ is an *1-cycle* because its boundary is 0.

Unoriented p-cycle group

Denoted by Z_p, this is the group of all p-cycles together. Since the boundary of the members of Z_p satisfies $\partial Z_p = 0$, then Z_p is a kernel of C_p. In the example above, Z_1 is comprised of only the cycle $c = \sigma_{\{v_0,v_1\}} \oplus \sigma_{\{v_1,v_2\}} \oplus \sigma_{\{v_0,v_2\}}$.

Unoriented p-boundary group

Denoted by B_p, this is the group of all p-chains that are the boundary of a $(p+1)$-chain. In the example above, the 1-chain $c = \sigma_{\{v_0,v_1\}} \oplus \sigma_{\{v_1,v_2\}} \oplus \sigma_{\{v_0,v_2\}}$ is the boundary of the 2-chain $\sigma_{\{v_0,v_1,v_2\}}$ so c belongs to group B_1.

Unoriented p-homology group

As in def. 1.95, the p-th homology group is the *quotient group* $\mathcal{H}_p = Z_p/B_p$. The 1-chain $c = \sigma_{\{v_0,v_1\}} \oplus \sigma_{\{v_1,v_2\}} \oplus \sigma_{\{v_0,v_2\}}$ from the previous example is the only member of Z_1 and B_1:

$$\mathcal{H}_1 = \frac{\sigma_{\{v_0,v_1\}} \oplus \sigma_{\{v_1,v_2\}} \oplus \sigma_{\{v_0,v_2\}}}{\sigma_{\{v_0,v_1\}} \oplus \sigma_{\{v_1,v_2\}} \oplus \sigma_{\{v_0,v_2\}}} = \emptyset$$

$\mathcal{H}_1 = \emptyset$ because it is the only coset left in $\sigma_{\{v_0,v_1\}} \oplus \sigma_{\{v_1,v_2\}} \oplus \sigma_{\{v_0,v_2\}}$.

Unoriented p-Betti number

The p-th Betti number is the rank of the p-th homology group $\beta_p = rank(\mathcal{H}_p)$ where the rank is the cardinality of the maximal independent subset. In the previous example, this is:

$$\beta_1 = rank(\mathcal{H}_1) = rank(\{\emptyset\}) = 0$$

By def. 1.96, β_1 provides information about the number of holes in the space. In the previous example, β_1 is computed from a 2-*simplex* which is a filled face that does not have holes thus the expected result.

1.6.3   Homology of abstract simplicial complexes

Definition 1.98 (Abstract p-chain) *A p-chain is a subset of p-simplices in a simplicial complex K.*

Example: Consider the 3-*simplex* solid tetrahedral (Fig. 1.2ivd); ignoring any geometrical considerations, the 3-*simplex* is defined (in a combinatorial form) as the pair $K = (V_K, S_K)$ where $V_k = \{v_1, v_2, v_3, v_4\}$ are vertices and S_K are simplices (faces) in K. All p-chains in K are shown below:

3-chains	$\{v_0, v_1, v_2, v_3\}$
2-chains	$\{v_0, v_1, v_2\}, \{v_0, v_1, v_3\}, \{v_1, v_2, v_3\}, \{v_0, v_2, v_3\}$
1-chains	$\{v_0, v_1\}, \{v_0, v_2\}, \{v_1, v_2\}, \{v_1, v_3\}, \{v_2, v_3\}, \{v_0, v_3\}$
0-chains	$\{v_0\}, \{v_1\}, \{v_2\}, \{v_3\}, \{\emptyset\}$

Definition 1.99 (Abstract p-chain group C_p) *This is a free Abelian group generated by p-chains in a simplicial complex K with the symmetric difference $\bigoplus$ as the operator:*

$$C_p = (p\text{-chain}, \bigoplus)$$

In this group, algebraic operations of p-chains can be performed.

Example: Consider the simplicial complex $K = (V_K, S_K)$ with $V_k = \{v_1, v_2, v_3, v_4\}$, and two 1-chains $\sigma_1 = \{\{v_1, v_2\}, \{v_1, v_3\}, \{v_2, v_3\}\}$ and $\sigma_2 = \{\{v_2, v_3\}, \{v_3, v_4\}, \{v_2, v_4\}\}$, then:

$$\sigma_1 \bigoplus \sigma_2 = \{\{v_1, v_2\}, \{v_1, v_3\}, \{v_2, v_3\}\} \bigoplus \{\{v_2, v_3\}, \{v_3, v_4\}, \{v_2, v_4\}\}$$
- $= \{\{v_1, v_2\}, \{v_1, v_3\}, \{v_2, v_3\}, \{v_3, v_4\}, \{v_2, v_4\}\} \setminus \{v_2, v_3\}\}$
- $= \{\{v_1, v_2\}, \{v_1, v_3\}, \{v_3, v_4\}, \{v_2, v_4\}\}$

This operation is shown by the geometric representation of K in Fig. 1.3v.

Definition 1.100 (Boundary of an abstract p-simplex) *For every p-simplex σ of a simplicial complex K, the boundary is defined by the alternating sum introduced in def. 1.92:*

$$\partial_p(\sigma) = \sum_{i=0}^{p} (v_0, \ldots, \widehat{v_i}, \ldots, v_p)$$

Intuitively, this is the set of the faces of $(p-1)$-*simplexes*.

Example: Consider a p-*simplex* with $p = 3$ (solid tetrahedral; Fig. 1.2ivd). Under the combinatorial representation of $K = (V_K, S_K)$ with $V_k = \{v_1, v_2, v_3, v_4\}$ vertices, the boundary of K is the set of faces of the $(p-1)$-*simplexes* which is the set of 2-*chains* in K:

$$\partial K = \{\{v_0, v_1, v_2\}, \{v_0, v_1, v_3\}, \{v_1, v_2, v_3\}, \{v_0, v_2, v_3\}\}$$

Definition 1.101 (Boundary of an abstract *p-chain* ∂_p) *This is the symmetric difference $\bigoplus$ of the boundary of the simplexes of an abstract p-chain.*

Example: Consider the 2-chain obtained above $c = \{\{v_0, v_1, v_2\}, \{v_0, v_1, v_3\}, \{v_1, v_2, v_3\}, \{v_0, v_2, v_3\}\}$, therefore the boundary is the following:

$$\partial c = \quad \partial\{v_0, v_1, v_2\} \bigoplus \partial\{v_0, v_1, v_3\} \bigoplus \partial\{v_1, v_2, v_3\} \bigoplus \partial\{v_0, v_2, v_3\}$$

Definition 1.102 (Abstract *p-cycles*) *A cycle c of dimension p is a p-chain with an empty boundary $\partial c = 0$.*

Example: The boundary of the previous 2-chain $\partial\{v_0, v_1, v_2\} \bigoplus \partial\{v_0, v_1, v_3\} \bigoplus \partial\{v_1, v_2, v_3\} \bigoplus \partial\{v_0, v_2, v_3\}$ is the following:

$$\begin{aligned}
\partial c = \quad & \{v_0, v_1\} \bigoplus \{v_0, v_2\} \bigoplus \{v_1, v_2\} \bigoplus \\
& \{v_0, v_1\} \bigoplus \{v_0, v_3\} \bigoplus \{v_1, v_3\} \bigoplus \\
& \{v_1, v_2\} \bigoplus \{v_1, v_3\} \bigoplus \{v_2, v_3\} \bigoplus \\
& \{v_0, v_2\} \bigoplus \{v_0, v_3\} \bigoplus \{v_2, v_3\} = \emptyset
\end{aligned}$$

The *2-chain* $c = \{\{v_0, v_1, v_2\}, \{v_0, v_1, v_3\}, \{v_1, v_2, v_3\}, \{v_0, v_2, v_3\}\}$ is a *2-cycle*.

Abstract *p*-cycle group

Denoted by Z_p, this is the group of all *p-cycles*. Since the boundary of the members of Z_p satisfies $\partial Z_p = 0$ then Z_p is a kernel of C_p. **Example**: Consider (section 1.6.2) the problem of obtaining the boundary of a *2-simplex* σ spanned by the vertices $\{v_0, v_1, v_2\}$ of the triangular face of Fig. 1.2ivc. In this case, Z_1 is comprised of the cycle $c = \sigma_{\{v_0, v_1\}} \bigoplus \sigma_{\{v_1, v_2\}} \bigoplus \sigma_{\{v_0, v_2\}}$.

Abstract *p*-boundary group

Denoted by B_p, this is the group of all *p-chains* that are boundaries of a *(p+1)-chain*. Considering the example above, the *1-chain* $c = \sigma_{\{v_0, v_1\}} \bigoplus \sigma_{\{v_1, v_2\}} \bigoplus \sigma_{\{v_0, v_2\}}$ is the boundary of the *2-chain* $\sigma_{\{v_0, v_1, v_2\}}$ ergo c belongs to group B_1.

Abstract p-homology group

As in def. 1.95, the p-th homology group is the *quotient group* $\mathcal{H}_p = Z_p/B_p$. The *1-chain* $c = \sigma_{\{v_0,v_1\}} \bigoplus \sigma_{\{v_1,v_2\}} \bigoplus \sigma_{\{v_0,v_2\}}$ from the previous example is the only member of Z_1 and B_1:

$$\mathcal{H}_1 = \frac{\sigma_{\{v_0,v_1\}} \bigoplus \sigma_{\{v_1,v_2\}} \bigoplus \sigma_{\{v_0,v_2\}}}{\sigma_{\{v_0,v_1\}} \bigoplus \sigma_{\{v_1,v_2\}} \bigoplus \sigma_{\{v_0,v_2\}}} = \emptyset$$

$\mathcal{H}_1 = \emptyset$ because it is the only coset left in $\sigma_{\{v_0,v_1\}} \bigoplus \sigma_{\{v_1,v_2\}} \bigoplus \sigma_{\{v_0,v_2\}}$.

Abstract p-Betti number

The p-th Betti number is the rank of the p-th homology group $\beta_p = rank(\mathcal{H}_p)$, where the rank is the cardinality of the maximal independent subset. In the example above, this is:

$$\beta_1 = rank(\mathcal{H}_1) = rank(\{\emptyset\}) = 0$$

By def. 1.96, β_1 provides information about the number of circular holes. In the example, $\beta_1 = 0$ because β_1 is obtained from a *2-simplex* which is a filled face with no holes.

1.7   CHAPTER FIGURES

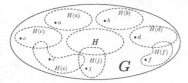

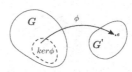

(i) Cosets generated by points $a,b,c,\ d,e,f,j \in G$ over the subgroup H.

(ii) A kernel of the homomorphism ϕ.

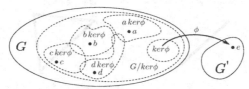

(iii) Example of quotient group $G/ker\phi$.

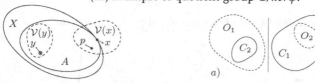

(iv) Example of limit and contact points.

(v) Example of open O_1, O_2 and closed C_1, C_2 sets.

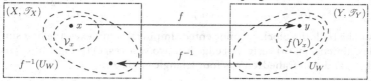

(vi) Example of a continuous mapping.

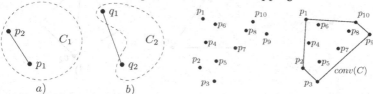

(vii) Example of: a) a convex set, and b) a non-convex set.

(viii) Set of points $C = \{p_1, ..., p_{10}\}$ and its convex hull $conv(C)$.

Figure 1.1: A diagrammatic representation of groups: fundamentals, mappings, and convex sets.

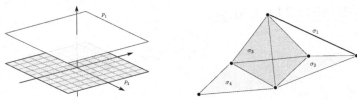

(i) Example of an affine vector space.

(ii) Example of a simplicial complex.

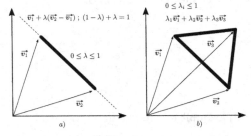

(iii) Example of k-simplexes generated by the convex hull.

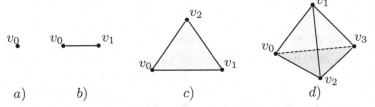

(iv) The building blocks of unoriented simplicial complexes; the first simplex dimensions: a) vertex, b) edge has two vertices, c) triangle has three edges, and d) tetrahedron has four triangles as faces.

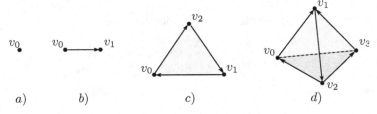

(v) Example of oriented simplexes in an affine space: pair of points span vectors oriented according to the listing order of their points. Building blocks of oriented geometric simplicial complexes: a) 0-simplex (vertex v_0), b) 1-simplex (oriented edge v_0, v_1), c) 2-simplex (oriented face v_0, v_1, v_2), and d) 3-simplex (oriented tetrahedron).

Figure 1.2: Simplicial complexes as affine spaces.

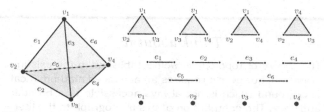

(i) Faces of a tetrahedron: one face in dimension 3 ($3-simplex$) has four faces of dimension 2 ($2-simplex$), six faces in dimension 1 ($1-simplex$), and four faces of dimension 0 ($0-simplex$) which are called vertices. The total number of proper faces is $15 = 2^4 - 1$ including $\{\emptyset\}$.

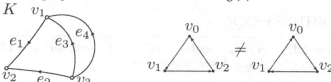

(ii) Example of an oriented complex K and two of its oriented 1-chains in K: $\{(e_1 + e_2 + e_3)(e_3 - e_4)\}$. An arbitrary combination produces an oriented 1-chain as $\{(e_1 + e_2 + 2e_3 - e_4)\}$.

(iii) The two ways of orienting a simplex; once the orientation is chosen it does not matter which vertex the list starts from because they are equivalent in the description of orientation.

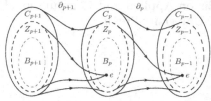

(iv) Abelian groups of the chain, cycle, and boundary under the mapping ∂. The cycle group Z_p and boundary subgroups B_p are shown as kernels and images of the boundary map ∂_p.

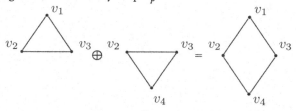

(v) Algebraic operations of $p-chains$.

Figure 1.3: Chain and boundary mappings.

Think about it...

The theory of topology is well-established in the existing literature; however it requires an advanced mathematical background which is not always accessible to a broader audience. The contribution of this monograph to the literature of computational topology is to provide a simplified path for accessing homology theory.

FURTHER READING

Dugundji, J. (1966). *Topology*. Allyn and Bacon series in advanced mathematics. Allyn and Bacon.

Ghrist, R. (2014). *Elementary Applied Topology*. Createspace Independent Pub.

Hilton, P. and Wylie, S. (1967). *Homology Theory: An Introduction to Algebraic Topology*. Cambridge University Press.

Zomorodian, A. (2005). *Topology for Computing*. Cambridge Monographs on Applied and Computational Mathematics. Cambridge University Press.

CHAPTER 2

Persistent homology of images

2.1   INTRODUCTION

The chapter presents a methodology for computing homology groups from digital images. The homology groups of a topological space provide intrinsic information about the number and type of holes in the space. This can be used to describe the structure of an image or for determining similarities between images.

Computing homology groups from images involves obtaining a configuration of simplicial complexes or other regular cell complexes that approximate the topology of the subjacent space. Existing literature shows that this is achieved by embedding the image in a Euclidean space where simplicial complexes are defined by considering image pixels as vertices. Then, higher dimensional simplices are spanned by an algorithm based on a distance function, e.g., alpha, Vietoris-Rips, Čech, etc. A significant disadvantage of this approach lies in the large amount of complexes spanned by the pixels, which can rapidly make it harder to compute the already computationally expensive Betti numbers.

To compute homology groups, the first task is to find a suitable model for representing a digital image. Existing approaches are frequently dependent on image contrast, e.g., the Fourier transform or wavelet decomposition. While this is not a problem in certain applications like image compression, it is inconvenient for tasks like image segmentation since invariance to contrast changes is highly sought-after. The elements of *Mathematical Morphology* (MM) are utilized to achieve such representations. MM is an established theory and technique for the analysis of geometric structures based on set theory, lattice theory, topology, and random functions [38]. MM deals with contrast invariant objects; the most basic ones are upper-level sets $[\phi \geq \lambda] = \{x, \phi(x) \geq \lambda\}$ or level lines $\partial[u \geq \lambda]$ of an image intensity function $\phi(x)$, where $x \in X$ is the spatial domain of its pixels. In the chapter, a new morphological operator is introduced for obtaining a decomposition of the digital image at different scales (a kind of data coarsening). This decomposition is used to build a finite approximation of the image based on abstract simplicial complexes which are suitable to compute homology groups.

The approach described here shares similarities with ref. [6] that studied morphological operators in digital images and the use of MM to perform tasks in image analysis. However, the difference is in the computed homology groups; the main contribution here is the formulation of a new method for spanning simplicial complexes from images.

2.2 DIGITAL IMAGES

A digital image is a discrete representation of objects in a visual field that have spatial and intensity information. This representation is set in a digital space in which each point can be defined as an integer vector of coordinates. The set $\mathbb{Z}^n$ is called the *n-dimensional* digital plane where the elements are called points. A non-empty subset of the *n-dimensional* digital plane is called the *n-dimensional digital set*. Here, the dimension is going to be restricted to two.

Definition 2.1 (Digital image) *If X is an n-dimensional set and $\mathcal{V}$ a given value space, then the function $\phi : X \to \mathcal{V}^n$ is called a digital image. The set X is the coordinate set while $\mathcal{V}$ is the value set of ϕ.*

Definition 2.2 (Pixel and binary image) *The ordered pair $(x, \phi(p))$ with $x \in X$ is called a pixel where x are the coordinates and $\phi(x)$ is the intensity of light waves falling on the optical sensor and recorded in the corresponding picture element. When the value space of ϕ is set on $\mathbb{Z}_2$ then the group spanned by the integers $\{0, 1\}$ is called* *a binary image. The foreground of the binary image is the set as* $P_1 = \{x \in X \mid \phi(x) = 1\}$ *and the background as* $P_0 = \{x \in X \mid \phi(x) = 0\}$. *Often, the foreground and background elements are called object and background points, respectively.*

Definition 2.3 (Grey-level image) *This is a digital image in which the value space of* ϕ *is set on* $\mathbb{Z}$ *or* $\mathbb{R}$. *A grey-level image is modelled as a topographical surface; the function* ϕ *is a surface* $I_m(x,y) = (x,y,\phi(x,y))$ *of the space (Fig. 2.1i).*

Definition 2.4 (Multichannel digital image) *A multichannel digital image is a grey-level digital image where the pixel at* (x,y) *is a* $1 \times n$ *array with n being the number of intensity channels recorded for each pixel. Often, each channel records intensities at different ranges of the light spectrum. An RGB color image is a multichannel digital image with* $n = 3$ *such that each element indicates the red, green, and blue intensities of the pixel.*

2.3 A NEIGHBORHOOD SYSTEM FOR IMAGES

Open sets are formulated in the space of grey-level images. There are two types of neighborhoods to consider: a **spatial-based neighborhood** defined by the spatial distribution of points and an **intensity-based neighborhood** defined in the range of the digital image function ϕ. This topic has been partially studied in refs. [5] and [34].

Recall from section 1.3 that a set X endowed with a distance function d spans a metric space and from that a topological space can be derived. $\mathbb{Z}^n$ is a space where it is possible to define a distance function, e.g., a Euclidean distance function $d_E : \mathbb{Z}^n \to \mathbb{R}$ such that:

$$d_E(x,y) = \sqrt{\sum_{i=1}^{m}(x_i - y_i)^2}$$

A simpler metric is the Taxicab metric defined by:

$$d_T(x,y) = \sum_{i=1}^{m}|x_i - y_i|$$

A set of points in a digital image with a distance function d is a metric space (X,d), hence a digital image I_m is in terms of topological spaces a finite metric space.

2.3.1 Spatial-based neighborhood

Definition 2.5 (Open neighborhood of a point) *In a similar fashion with the definition of open balls in a metric space, the **open neighborhood of a point** p in a digital image X ($p \in X$) with a metric d is defined by:*

$$N_p = \{x \in X : d(p,x) < \varepsilon\}, \varepsilon > 0$$

N_p does not contain *border points* but contains all the points that are sufficiently near p with $d(x,p) < \varepsilon$.

Example: For $\varepsilon = 1$ the neighborhood of a point $p \in X$ is a set that contains p and none of its surrounding points (Fig. 2.1ii).

Definition 2.6 (Open set of points) *Given a set U in the space of image points X ($U \subset X$) then U is an open set if all points $x \in U$ have a neighborhood N_x contained in U.*

Definition 2.7 (Interior point) *Given a set of image points X and $A \subset X$, then a point $x \in X$ is an **interior point** of A if $x \in A$ and there is a neighborhood N_x of x that is a subset of A ($N_x \subset A$):*

$$I_A = \{x \in A : \exists N_x \subset A\}$$

Definition 2.8 (Boundary point) *A boundary point is a point x of a set A in the space of image points X ($A \subset X$) such that any neighborhood N_x of x intersects A and its complement A^c ($A^c = X \setminus A$):*

$$B_A = \{x : x \in A, A \subset X, \forall N_x \, N_x \cap A \neq \emptyset, N_x \cap A^c \neq \emptyset\}$$

Definition 2.9 (Closed neighborhood of a point) *The **closed neighborhood of a point** p in a digital image X with a metric d is defined by:*

$$N_p = \{x \in X : d(p,x) \leq \varepsilon, 0 < \varepsilon\}$$

Here, N_p contains the *border points*.

2.3.2 Intensity-based neighborhood

Let X be a set of points in a digital image where $x \in X$ is a point and ϕ is the digital image function. The neighborhood of x induced by ϕ is denoted by $N_{\phi(x)}$. Three different types of image point neighborhoods are introduced.

Definition 2.10 (Unbounded neighborhood) *This is a system of points with values sufficiently close to ϕ based on pixel intensity; the points can be anywhere in the image:*

$$N_{\phi(x)} = \{y \in X : d(\phi(x), \phi(y)) < \delta\}, \delta > 0$$

Definition 2.11 (Bounded neighborhood) *There is a restriction on the location of an image point in relation to the center of the neighborhood:*

$$N_{\phi(x),\varepsilon} = \{y \in X : d(\phi(x), \phi(y)) < \delta \wedge d(x,y) < \varepsilon\}, \delta > 0$$

Definition 2.12 (Identically bounded neighborhood) *This contains points that are indistinguishable in terms of intensity. Every point y matches the intensity value of point x:*

$$N_{\phi(x),\varepsilon} = \{y \in X : d(\phi(x), \phi(y)) = 0 \wedge d(x,y) < \varepsilon\}, \delta > 0$$

The digital plane

Here, the aim is to provide a definition of the concept of connected components [13]. A simple model of the digital plane is to treat the set of all points in $\mathbb{R}^2$ as having integer coordinates. The concepts of digital topology and geometry are concerned with topological and geometrical properties of subsets of the digital plane (digital sets). In the context of image processing, the digital plane is a mathematical model of black and white elements; the set of black points S and the complement set of white points $\mathbb{C}S$. Given a point $P = (m,n)^\top \in \mathbb{Z}^2$ then the 8-neighbors of P are points with integer coordinates $(k,l)^\top$ such that max $(|m-k|, |n-l|) \leq 1$. A practical notation for the 8-neighbors of P is shown in Table 2.1.

Table 2.1: The 8-neighbor labels of pixel P. Neighbors with an even and odd number are direct (4-neighbors of P) and indirect neighbors, respectively. The 8-neighborhood $\mathcal{N}_8(P)$ of P is the set of all 8-neighbors of P (excluding P). The 4-neighborhood $\mathcal{N}_4(P)$ of P is the set of all 4-neighbors of P.

row	column $n-1$	n	$n+1$
$m-1$	$N_3(P)$	$N_2(P)$	$N_1(P)$
m	$N_4(P)$	P	$N_0(P)$
$m+1$	$N_5(P)$	$N_6(P)$	$N_7(P)$

Let κ be any of the numbers 4 or 8 and let $I = \{0, 1, \ldots, n\}$ be a finite interval of consecutive integers. A digital κ-path or simply a path $\mathcal{P}$ is a sequence of points $\{P_i\}_{i \in I}$ in $\mathbb{Z}^2$ such that P_i and P_j are κ-neighbors of each other when $| i - j |= 1$. The order induced by the numbering of the points in the path is essential. For $P \in \mathcal{P}$, the κ-*degree* or degree of P with respect to $\mathcal{P}$ is defined by the number $| \mathcal{P} \cap \mathcal{N}_k(P) |$. A point of $\mathcal{P}$ having degree 1 is called an end point. An immediate consequence of the definition is that any point in the path has degree of at least one. Hence, at most two end points exist in the path. End points can only correspond to numbers 0 or n. A path with the property $P_0 = P_n$ is called a closed path which contains no end points. A κ-*path* $\mathcal{P}$ is called a κ-arc or simply an arc if it has the additional property that for any two points P_i, $P_j \in \mathcal{P}$ that are not end points: $P_i \in \mathcal{N}_\kappa(P_j)$ implying $| i - j |\le 1$. Consequently, an arc is a path which does not intersect with itself with the possible exception of its end points. Any point in an arc has order one or two.

Lemma 2.1 *Let $\mathcal{P}$ be a path with two end points, then an arc P_0 exists which is completely contained in $\mathcal{P}$ and has the same end points.*

Proof: If $\mathcal{P}$ is a κ-path but not a κ-arc then it contains at least one pair of points P_i and P_k such that $P_i \in \mathcal{N}_\kappa(P_k)$ with $| i - k |> 1$. Assuming without loss of generality that $k > i$, then the path:

$$P' := \{P_0, P_1, \cdots, P_{i-1}, P_i, P_k, P_{k+1}, \cdots, P_n\}$$

is contained in $\mathcal{P}$ and has the end points P_0 and P_n but has fewer elements than $\mathcal{P}$. When the procedure is repeated then an arc is obtained with the desired property. A digital set $S \subseteq \mathbb{Z}^2$ is called a κ-connected set when for any two points $P, Q \in S$ a path or more precisely an arc $\mathcal{P}$ exists that is completely contained in S and that contains both P and Q. A connected component of a set $S \subseteq \mathbb{Z}^2$ is the maximal subset of S which is connected. This is a purely graph theoretical approach of connectedness and it is fundamental for analyzing digital sets. Through connectedness, some rudimentary topology in $\mathbb{Z}^2$ is introduced therefore the familiar terminology about 4- and 8-topology for the digital plane is adopted.

2.4 COMPUTING HOMOLOGY FROM IMAGES

Natural and medical images are rich in spatially distributed patterns of intensity, hence spatial or intensity neighborhoods (or both) can be defined. The challenge is to take advantage of this in a way that allows the computation of homology groups. The chapter introduces a procedure for computing the homology of grey-level images through computing the homology of the space obtained when the image is fragmented at different scales. The notion of fragmenting an original space to access its homology coincides with ref. [35] in that the coarsening of spaces is a fundamental task for studying their homology.

2.4.1 The digital image function ϕ

A grey-level image can be viewed as a topographic surface. A standard way to represent the topography on a map is to draw *level lines*; these are lines that are produced from connecting points of equal elevation in respect to a surface function. Level lines do not cross, divide or split, and they are closely spaced to represent steep slopes. Conversely level lines that are spaced far apart represent gentle slopes (Fig. 2.1v). Although a dense set of level lines provides an accurate geometric representation, the focus here is mainly on the inclusive relationships between adjacent level lines.

Let's consider a 2D grey-level digital image as a parameterized 3D surface $\mathcal{S}(i,j) = (i,j,\phi(i,j))$ where the first two coordinates (i,j) are the spatial domain of the image $i,j \in X \subset \mathbb{Z}^2$ and the third is the value of $\phi : \mathbb{Z} \times \mathbb{Z} \to \mathbb{R}$ representing pixel intensity at (i,j). The convention adopted here is that for an n-dimensional coordinate system, the first $n-1$ coordinates constitute the spatial domain of $\mathcal{S}$ and the nth-coordinate is the value domain of the surface. Under this convention, a grey-level image is a coordinate system of dimension $n=3$.

For grey-level images, ref. [32] shows two ways for representing the digital image function ϕ as sets: the cross-sectional [39] and the umbra [44] approaches. The cross-sectional representation uses multiple thresholds to obtain binary images; formulation and implementation are cumbersome since multiple structural elements and operators over the stack are required. The umbra approach is simpler to implement and intuitive; the umbra is the set of all points contained below a given surface value. The utilized morphological operator is derived from the *umbra* therefore the concept of a *top surface* is defined first and then extended for defining the umbra and the operator.

Definition 2.13 (Top surface of a digital image) *Let $S \subseteq (\mathbb{Z}^2 \times \mathbb{R})$ and $X = \{(i, j) \in \mathbb{Z}^2 \mid$ for some $\phi(i, j) \in \mathbb{R}, (i, j, \phi(x, y)) \in S\}$, then the top surface of S denoted by $T[S] : X \to \mathbb{R}$ is defined by:*

$$T[S](i, j) = max\{\phi(i, j) \mid (i, j, \phi(i, j)) \in S\}$$

The *top surface* of a surface S is a function defined on the projection of S to $(n-1)$ coordinates. As an example, consider a closed curve A where $n=2$ then $F = \{x \in \mathbb{E}^{n-1} \mid$ for some $y \in \mathbb{E}, (x, y) \in A\}$. For each $(n-1)$-tuple x, the *top surface* of A at x is the highest value of y such that the n-tuple $(x, y) \in A$ is $T[A](x) = max\{y \mid (x, y) \in A\}$ (Fig. 2.1iii). If the space of A is Euclidean then the *top surface* can be expressed in terms of the *supremum*; otherwise if the space is discrete the concept of *maximum* is utilized.

Definition 2.14 (Umbra) *A set of points $A \subseteq (\mathbb{Z}^2 \times \mathbb{R})$ is an umbra iff $(i, j, \epsilon_1) \in A$ implying that $(i, j, \epsilon_2) \in A$ for every $\epsilon_1 \leq \epsilon_2$. The umbra of S is the set consisting of the surface S and everything below it. Let $X \subseteq \mathbb{Z}^2$ and $\phi : \mathbb{Z}^2 \to \mathbb{R}$, then the umbra of ϕ denoted by $U[\phi] \subseteq \mathbb{Z}^2 \times \mathbb{R}$ is defined by (Fig. 2.1iv):*

$$U[\phi] = \{(i, j, \epsilon) \in X \times \mathbb{R} \mid \epsilon \leq \phi(i, j)\}$$

2.4.2 Projected umbra

The motivation of including the *top* and *umbra* sets is to utilize them for defining a special projection that is called *projected umbra*. The *projected umbra* is an umbra constructed by slicing the function ϕ at different heights and projecting the umbra at a given height. This projection is comprised of points $(i, j, \phi(i, j))$ such that the value of the digital image function $\phi(i, j)$ is greater than or equal to a given value h (the formal definition is attributed to Prof. Dr. Luis Salinas Carrasco). Let's consider a discrete set of points on a continuous surface $S \subset (\mathbb{Z}^2 \times \mathbb{R})$ (local 2-dimensional manifold). Let's denote a discrete finite set of points $\mathcal{R}$ on S and points (x_i, y_i, z_i) with $i = 1 \ldots n$ such that:

$$\mathcal{R} = \{(x_i, y_i, z_i) \in S : i = 1, \ldots, n\} \subset S \subset \mathbb{Z}^2 \times \mathbb{R}$$

First, a projection is defined by:

$$\pi_h = \mathcal{R} \to \mathbb{Z}^2 \times \mathbb{R}$$

such that $(h \in \mathbb{R})$:

$$(x_i, y_i, z_i) \xrightarrow{\pi_h} \pi_h(x_i, y_i, z_i) = \begin{cases} ((x_i, y_i), h) & , if \ z_i \geq h, \\ \emptyset & , if \ z_i < h \end{cases}$$

The definition is almost complete but a problem arises regarding $\emptyset$. A function to a space $(\mathbb{Z}^2 \times \mathbb{R})$ must have a well-defined element of $(\mathbb{Z}^2 \times \mathbb{R})$ for every point $(x_i, y_i, z_i) \in \mathcal{R}$, therefore $(0, 0, 0)$ is assigned arbitrarily instead of $\emptyset$.

Definition 2.15 (Projected umbra) *The projected umbra $U_{\pi_h}[\phi]$ of a surface described by ϕ and a scalar value h is defined by:*

$$U_{\pi_h}[\phi] \equiv U_{\pi_h}[\phi(x_i, y_i)] := \{\pi_h(x_i, y_i, \phi(x_i, y_i)) : i = 1, \dots, n\},$$

where:

$$\pi_h(x_i, y_i, \phi(x_i, y_i)) = \begin{cases} ((x_i, y_i), h) & , if \ \phi(x_i, y_i) \geq h, \\ (0, 0, 0) & , if \ \phi(x_i, y_i) < h \end{cases}$$

The **projected umbra** produces a (geographic) chart of all points on $\mathcal{R}$ above h. Given a fixed scalar h then the morphology of such a set of points can be analyzed, e.g., to obtain its connected components (Fig. 2.2ii). This definition shares similarities with the *binary umbra* introduced in ref. [31], however the *projected umbra* defines a 4D object (x_i, y_i, z_i, h) such that for a given value h a 3D slice can be obtained. The *binary umbra* obtains morphological features from binary level set images. The similarity between them is that the original image can be reconstructed through stacking its umbras at different level sets.

Definition 2.16 (Hierarchical set of umbra projections) *Let's consider the range of the digital image function $\phi : X \to [\Gamma_0 \dots \Gamma_1]$ with $[\Gamma_0 \dots \Gamma_1] \in \mathbb{R}$, $\Gamma_0 < \Gamma_1$ starting from its regional extrema (local minima and maxima in ϕ). A strictly ordered set of real numbers $H_F = \{h_0, h_1, h_2, \dots, h_n\}$ can be defined (n is an arbitrary natural number) such that:*

$$\Gamma_0 \leq h_0 < h_1 < h_2 < \dots < h_n \leq \Gamma_1 \quad where \ h_0, h_1, h_2, \dots, h_n \in \mathbb{R}$$

The projected umbra $U_{\pi_h}[\phi]$ with h taking values from H_F produces a sequence called the hierarchical set of umbra projections $U_\pi[\phi]$. This constitutes a kind of foliation of the image space (Fig. 2.2iii). Decomposing the image into a collection of disconnected sets of pixels is the process of image dismantling. Each disconnected set of pixels becomes an equivalence class for its members; each disjointed component is considered as a single abstract object. This perspective allows to represent the image as a hierarchical tree structure.

2.4.3 Inclusion tree of umbra projections

Here, the concept of a tree structure is introduced for handling the collection of projections of an image and mathematical descriptions are provided. The hierarchical set of umbra projections $U_\pi[\phi]$ defines a four-dimensional object $(x_i, y_i, \phi(x_i, y_i), h_n)$ constructed by taking values of $h_n \in H_F$ starting from the local minima and maxima in the digital image function. From a given h_0, it is possible to obtain a 3D set of points $p_i = (x_i, y_i, h_0)$ where $i = 1 \dots m$ is the pixel index in the spatial domain X of the image. The spatial domain $x_i, y_i \in X$ is the same for any chosen value of h_n.

Set of pixels in the projection

The set of pixels in the projected umbra $U_{\pi_h}[\phi]$ (for $h = h_n$) is defined by:

$$\mathcal{P}_n = \{(x_i, y_i) \in X : (0,0,0) \neq (x_i, y_i, h_n) \in U_{\pi_{h_n}}[\phi]\} \quad h_n \in H_F$$

$\mathcal{P}_n$ is the set of pixel coordinates in the projection $U_{\pi_{h_n}}$ for which $\phi(x_i, y_i) \geq h_n \in range\ of\ \phi$. Two points $p_1, p_2 \in \mathcal{P}_n$ are connected if p_1 lies in an open neighborhood of p_2 or vice-versa. A connected set of points $\mathcal{C}$ is a set which cannot be partitioned into two nonempty subsets such that each subset has no points in common with the set closure of the other. The maximal connected subsets in $\mathcal{P}_n$ are called the connected components $\mathcal{CC}(\mathcal{P}_n)$ of $\mathcal{P}_n$ inducing a partition in X; for $c_1, c_2 \in \mathcal{P}_n$ their arbitrary intersection is empty $c_i \cap c_j = \emptyset, i \neq j$. The focus here is on neighborhoods of 8-adjacent pixels $\mathcal{N}_8$, however in order to build the formulation the neighborhood concept is kept as general as possible. Let $\mathcal{CC}$

be the set of all connected components in the set $\mathcal{P}_k$ (pixel coordinates) in umbra $U_{\pi_{h_n}}[\phi]$ for all height values h_k $(k = 1,\ldots,n)$. CC is a summary of connected components for all height values h_k $(k = 1,\ldots,n)$. The connected components change for different k which is related to the *height* h_k of the plane giving rise to $U_{\pi_{h_k}}[\phi]$. Thus, for connected components c_k in $U_{\pi_{h_k}}[\phi]$ and c_l in $U_{\pi_{h_l}}[\phi]$ either $c_k \cap c_l = \emptyset$ or $c_l \subseteq c_k$. In this sense, CC is a tree or the horizontal cuts at heights h_k of a tree. The notion of trees formed by subsets has been studied in ref. [6]. Here, only the fundamental theory blocks are shown required to describe the method of computing homology groups.

Definition 2.17 (Tree of subsets T) *A collection of subsets* T *of X is called a tree of subsets of X if it satisfies two properties:*

  i) T *contains X, and*

  ii) $C, D \in$ T *with either* $C \cap D = \emptyset$, $C \subseteq D$, *or* $D \subseteq C$; *in the last two cases C and D are called nested.*

The elements of the tree are called nodes. Thanks to property *(ii)*, there is no cycle in the tree.

Example: The domain of all connected components in $\mathcal{P}_k$ $(k = 1 \ldots n)$ is denoted by CC. Any two connected components c_i, c_j, $i \neq j$ in CC can be organized as a disjointed set $c_i \cap c_j = \emptyset$ or as nested sets $c_i \subseteq c_j$ or $c_j \subseteq c_i$. Hence, CC is a tree.

Definition 2.18 (Interval of a tree) *Let* $c_o \subseteq c_p \subseteq CC$ *then* $[c_o, c_p]$ *is defined as the interval of* T *between* c_o *and* c_p:

$$[c_o, c_p] = \{S : S \in \mathsf{T}, c_o \subseteq S \subseteq c_p\}$$

Definition 2.19 (Inf and Sup) *Given* $c_o \subseteq c_p \subseteq CC$ *then:*

$$inf[c_o, c_p] = \bigcap_{S \in [c_o, c_p]} S$$

$$sup[c_o, c_p] = \bigcup_{S \in [c_o, c_p]} S$$

Definition 2.20 (Bifurcation) $c_p \subseteq CC$ *contains a bifurcation in* T *if* $n_a, n_b \subseteq$ T *exists such that* $n_a, n_b \subseteq c_p$ *and* $n_a \cap n_b = \emptyset$. *Let* $c_o \subseteq c_p \subseteq CC$ *then there is a bifurcation between* c_o *and* c_p *if* $inf[c_o, c_p] \neq \emptyset$ *and there is a* $S \in T$ *such that* $S \subseteq c_p$ *and* $S \cap inf[c_o, c_p] = \emptyset$. *For example, in Fig. 2.3i the nodes* (cc_3, cc_4, cc_{14}) *contain a bifurcation.*

Definition 2.21 (Branch) *Let $c_o \subseteq c_p \subseteq CC$ then $[c_o, c_p]$ is a branch of* T *if there is no bifurcation between c_o and c_p. A branch $[c_o, c_p]$ contains a point $x \in X$ if there is a node $c_q \in [c_o, c_p]$ such that $x \in c_q$.*

Definition 2.22 (Upper branch) *Given a point $x \in X$ then S_x is defined as the smallest limit node containing x:*

$$S_x = inf[\{x\}, X] = \bigcap_{S \in T, x \in S} S$$

The upper branch at x is the set:

$$B_x = \{S : S \in T, x \in S, \text{ and } [S_x, S] \text{ is a branch}\}$$

Definition 2.23 (Leaf) *A leaf of* T *or simply a leaf is defined as any limit node $L = inf[c_o, c_p]$ containing no other shape. For example, in Fig. 2.3i the nodes $\{cc_{12}, cc_{23}, cc_{29}, cc_{27}, cc_{22}\}$ are leafs.*

2.4.4 Algorithms for computing homology

Given a 2D grey-level image (Fig. 2.1v) then the digital image function is defined by $\phi : \mathbb{Z}^2 \to \mathbb{R}$ (reminder: if the set of real numbers $\mathbb{R}$ is complete then it is ordered). Since the co-domain of ϕ is ordered, an ordered set of level lines on ϕ can be obtained by taking values at $\{l_0, \ldots, l_i\}$ such that $l_0 \le l_1 \le \ldots \le l_i$ with all level lines having real values in the range of ϕ (Fig. 2.2i). Algorithm 1 shows the pseudocode for this step.

Algorithm 1 Projected umbra

```
1: procedure PROJECTEDUMBRA(φ,l)                    ▷ φ : 2D digital image function
2:     U = BinaryArray(row,col) = false
3:     for each row in φ do
4:         for each col in φ do
5:             if φ(row,col) ≥ l then
6:                 U(row,col)= true;
7:     return U                                      ▷ U : 2D binary image
```

The image can be represented by a collection of projections at values $H_f = \{l_0, \ldots, l_i\}$. Connected components[1] at each level line l_i, namely $CC(l_i)$, can be obtained as in section 2.4.3. Algorithm 2 shows the pseudocode for obtaining and labeling connected components as described in ref. [37].

---
[1] A connected component is a set of points in the image induced by a neighborhood system that relates points in the set.

Algorithm 2 Labeling connected components

```
1: procedure LCONCOMP(imdata)                        ▷ imdata : 2D binary image
2:     linked = []
3:     [imrow,imcol] ← size(imdata);
4:     labels = integerarray (imrow,imcol);
5:     for row in imdata:
6:         for column in row:
7:             if data[row][column] is not Background
8:                 neighbors = connected elements with the current element's value
9:                 if neighbors is empty
10:                linked[NextLabel] = set containing NextLabel
11:                labels[row][column] = NextLabel
12:                NextLabel += 1
13:            else
14:     Find the smallest label
15:            L = neighbors labels
16:            labels[row][column] = min(L)
17:            for label in L
18:            linked[label] = union(linked[label], L)
19:     Second pass
20:        for row in data
21:            for column in row
22:                if data[row][column] is not Background
23:                labels[row][column] = find(labels[row][column])
24:     return labels
```

Algorithm 3 Inclusion tree

```
1: procedure INCLUSIONTREE(φ,H_f)                    ▷ φ : 2D digital image function
2:     tmp = array(m,n) = 0
3:     InclusionTree = null
4:     for each l in H_f do
5:         U_{π_l}[φ] = UmbraProjected (φ,l)
6:         CC_l       = lconcomp(U_{π_l}[φ])
7:         for each c_l in CC_l do
8:             if tmp(c_l) ≠ 0 then
9:                 InclusionTree (c_l) = tmp(c_l)
10:            else
11:                InclusionTree (c_l) = tmp(c_l)
12:                tmp(c_l) = c_l;
13:     return InclusionTree
```

By varying the height of the slicing plane from l_0 to l_i in the range of ϕ, connected components gradually split and disappear according to the morphology of the elevation surface (Fig. 2.3i). The progressive changes of connected components explain the inclusion relations on which the tree is constructed.

Connected components induced by the projection of umbras U_{π_m}, U_{π_n} located at two successive level lines l_m, l_n, respectively, span inclusion relationships [6]. Any two components $cc_i, cc_j \in \{U_{pi_m}, U_{pi_n}\}$ are either nested ($cc_i \subseteq cc_j$ or $cc_j \subseteq cc_i$) or disjointed ($cc_i \cap cc_j = \emptyset$). Such inclusion relations have a natural representation in the form of hierarchical structures. The hierarchical tree representing elements of CC and their inclusion relations is called an *inclusion tree*. In this representation, the root node is a unique connected component cc_0 that covers the entire image. The cc_0 matches a *level line* at the lowest value of ϕ. Conversely, leaves of the tree are local maxima of ϕ. Algorithm 3 shows the pseudocode for obtaining a tree structure from a grey-level image.

2.4.5 Construction of abstract simplicial complexes

The leaf nodes of a tree are likely to correspond to small disjointed areas in the image. Despite their small size, they are local maxima, therefore salient regions of ϕ. Leaf nodes are considered representatives of the branch they belong to. A bifurcation is where two or more leaf nodes join and the root of the tree is the union of all leaf nodes (Fig. 2.3ii). Algorithm 4 considers leaf nodes as representatives of their branches. Here, all intermediate nodes are progressively removed from the tree until only leaf nodes, bifurcation nodes, and the tree root remains. Nodes retain information about the height of the original level line.

Algorithm 4 Tree simplification

```
 1: procedure TREESIMPLIFICATION(T)                          ▷ T : Inclusion tree
 2:    L = FindTreeLeaves(T)
 3:    for each l in L do
 4:       link → get parent node(l, T)
 5:       node → l
 6:       while link ≠ 0 do
 7:          if number of sons(link, T) > 1
 8:          then                          ▷ parent node is a joint of several branches
 9:             set parent(node, T) = link
10:          else       ▷ parent node is an intermedia node that can be deleted.
11:             set parent(node, T) = get parent node(link, T)
12:             delete node(link, T)
13:          endif
14:          link = get parent node(node, T)
15:       endwhile
16:    endfor
17:    return T
```

Traversing the tree structure (Fig. 2.3ii) from the leaf nodes $\mathcal{L} = \{cc_{12}, cc_{23}, cc_{29}, cc_{27}, cc_{22}\}$ to the root produces several configurations of sets and subsets of leaf nodes. A graph is a natural representation of this result; the leaf nodes of the tree are graph vertices and the induced sets represent how vertices connect to each other. The first task consists of building simplicial complexes from the graph. By taking the set $\mathcal{L}$ of leaf nodes in the inclusion tree as the vertices of a simplicial complex K and the set S_K of the non-empty finite subsets of elements of $\mathcal{L}$ as the *simplices* within K, then:

a. $(\forall v \in \mathcal{L}), \{v\} \in S_k$

b. $(\forall \sigma \in S_k), (\forall \sigma' \subset \sigma, \sigma' \neq \emptyset), \sigma' \in S_k$

The abstract simplicial complex K is defined by the pair $K = (\mathcal{L}, S_k)$.

Example: For the set of leaf nodes $\mathcal{L} = \{cc_{12}, cc_{23}, cc_{29}, cc_{27}, cc_{22}\}$ (Fig. 2.3i) and the set of subsets $S_{l_5} = \{\{cc_{12}\}, \{cc_{23}\}, \{cc_{29}\}, \{cc_{27}, cc_{22}\}\}$ (Table 2.3ii), let's assess if the pair $K = (\mathcal{L}, S_{l_5})$ satisfies the axioms of a simplicial complex:

  a. $(\forall\, v \in \mathcal{L}),\ \{v\} \in S_k$;
  For a given $v = \{cc_{27}\}$, the condition $\{v\} \in S_k$ is not satisfied ($\{cc_{27}\} \neq S_{l_5}$) therefore K is not a simplicial complex

Due to the way that S_{l_i} is obtained (Table 2.3ii), it may not be possible to have all the required elements in order to form an abstract simplicial complex. Hence, an alternative method of construction is introduced.

For a finite set of subsets S_{l_h} of a set $\mathcal{L}$ then:

  1. the set of vertices V_{l_h} for all elements v_i in $\mathcal{L}$ such that ($\sigma \in S_{l_h}$) is either $\sigma = v_i \in \mathcal{L}$ or $\forall\, v_i \in \sigma,\, v_i \in \mathcal{L}$

  2. the pair $K = (V_{l_h}, S_{l_h})$ is an abstract simplicial complex

The *0-simplices* of K are points in V_{l_h}. A *n-simplex* has $n + 1$ points in $\sigma = [v_{i_0}, \ldots, v_{i_n}]$ where $v_{i_0} \ldots v_{i_n} \in V_{l_h}$.

Algorithm 5 presents the process for building simplicial complexes from the simplified inclusion tree.

Algorithm 5 Create simplicial sets

```
 1: procedure CREATESIMPLICIALSETS(T)                    ▷ T : Inclusion tree
 2:     H = HeightOfTheTree(T)
 3:     for h ∈ {1...H} do              ▷ height from leaves (1) to the root(H)
 4:         P = getAllNodesAtHeight(h, T)
 5:         S_h = {∅}
 6:         for each node ∈ P do
 7:             S_h = S_h ⋃ getNodeSons(node)          ▷ S_node is a set
 8:         endfor
 9:     endfor
                H
10:     S = ⋃ {S_h}
               i=1
11:     return S
```

Example: Repeating the previous example for $\mathcal{L} = \{cc_{12}, cc_{23}, cc_{29}, cc_{27}, cc_{22}\}$ and $S_{l_5} = \{\{cc_{12}\}, \{cc_{23}\}, \{cc_{29}\}, \{cc_{27}, cc_{22}\}\}$:

1. The set V_{l_5} has the elements of $\mathcal{L}$ required to span the set $\mathcal{S}_{l_5}$; $V_{l_5} = \{cc_{12}, cc_{22}, cc_{23}, cc_{27}, cc_{29}\}$

2. It is easy to verify that $K = (V_{l_5}, \mathcal{S}_{l_5})$ satisfies the axioms of simplicial complexes

2.5 PERSISTENT HOMOLOGY

The aim of persistent homology is to measure the lifetime of certain topological properties of a simplicial complex when simplices are added to the complex or removed from it [23]. The concept has been introduced in ref. [14]. Initial applications included the analysis of the persistence of topological features over a long range of parameters in point-cloud data. F_S is defined as the collection of all abstract simplicial complexes spanned by the inclusion tree levels obtained from a digital image. The aim is to use these complexes as a new representation of the image. A question that arises is pertinent to the elements of F_S that should be considered as more representative; instead of choosing some elements all of them can be utilized. Here, the focus is on the inclusion tree inducing an inclusion relation between complexes in F_S thus providing valuable information for characterizing a digital image. In addition, a terse but formal definition of persistent homology is provided.

2.5.1 Filtration

Using the method described in section 2.4.5, all abstract simplicial complexes are generated from the inclusion tree of Fig. 2.3ii. Complexes are represented as unoriented geometric simplicial complexes (V_{l_h} vertices and S_{l_h} subsets of V_{l_h}). A family of abstract simplicial complexes $F_S = \{K_{l_{12}}, K_{l_{11}}, K_{l_{10}}, \ldots, K_{l_0}\}$ can be obtained; simplices are always added but never removed implying a partial order which is called a filtration.

Definition 2.24 (Filtration) *A filtration of a complex K is a nested sequence of subcomplexes:*

$$\emptyset = K_0 \subseteq K_1 \subseteq K_2 \subseteq \ldots \subseteq K_n = K$$

A complex K with a filtration is called a filtered complex. A complex K is filtered by a filtration $\{K_i\}$, $i = 0, \ldots, n$ if $K_n = K$ and K_i is a subcomplex of K_{i+1} (for each $i = 0, \ldots, n-1$). Note that

$K_{i+1} = K_i \cup \sigma_i$ where σ_i is a set of simplexes. The sets σ_i provide a partial order on the simplexes of K. Filtered complexes arise naturally in many situations. In this case, complex K is filtered by providing an order $\sigma_0, \sigma_1, ..., \sigma_m$ to its simplexes according to the sequence of complexes induced from the inclusion tree and the sequence of its subcomplexes such that $K_i = \{\sigma_j \in K \mid 0 \le j \le i\}$.

The ordering of K_i is not necessarily the same order of the level lines in the inclusion tree (section 2.4.4). The ordering of K_i defines the entry sequence of each simplex during the filtration. Consider as an example the box l_6 of Fig. 2.4i; two simplexes appear: vertex e and edge $\{b, d\}$. Although it cannot be deduced which of them appears first, an arbitrary order of appearance for K has to be provided. Using the abstract complexes induced by the level lines as *filtration snapshots*, the complex grows from $K_0 = \{\sigma_0\}$ adding each simplex one at a time. It is assumed that if σ_i is a face of σ_j then σ_i enters the filtration before σ_j. Regarding the dimension p of the homology groups denoted by $\mathcal{H}_p$, if the step from $\mathcal{H}_p(K_i)$ to $\mathcal{H}_p(K_{i+1})$ is considered then several changes can occur; new homology classes can be created or existing homology classes can merge or become trivial. The purpose of persistent homology is to record these changes throughout the filtration.

2.5.2 Persistence

Given a filtered complex, the *ith* complex K_i has the associated boundary operators ∂_p^i and groups C_p^i, Z_p^i, B_p^i, and $\mathcal{H}_p^i$ (for all $i, p \ge 0$). The superscript i indicates the filtration index and the subscript p refers to the dimension of boundary maps, chains, cycles, and homology groups.

Definition 2.25 (n-persistent homology group) *The n-persistent homology group of dimension p (p-th homology group) of the K_i filtered complex refers to the homology in the filtration interval $K_i \subseteq ... \subseteq K_{i+n}$ denoted by:*

$$\mathcal{H}_p^{i,n} = Z_p^i / (B_p^{i+n} \bigcap Z_p^i)$$

This definition of homology is slightly different from the one given in def. 1.95. The meaning of $(B_p^{i+n} \bigcap Z_p^i)$ is that n steps after i changes may have occurred in homology classes, e.g., the entry of new classes. For the quotient to make sense, one can compute the *modulo* by the cycles in B_p^{i+n} that exist at Z_p^i. Since B_p^{i+n} and Z_p^i are both subgroups of C_p^{i+n} then $B_p^{i+n} \cap Z_p^i$ is also a subgroup of Z_p^i hence $\mathcal{H}_p^{i,n}$ is well-defined. The *n-persistent* Betti number (with dimension p) of complex K_i denoted by $\beta_p^{i,n}$ is the rank of $\mathcal{H}_p^{i,n}$. The **persistence** of a *p-cycle* created at the filtration index i and destroyed at time $j = i + n$ is $j - i$. This approach is based on a sequence of filtration indices known as **time-based persistence** resembling the time of birth and death of cycles during the filtration.

Definition 2.26 (Persistence complex) *A persistence complex C is a family of chain complexes $\{C_*^i\}_{i \geq 0}$ (* is a placeholder for dimension) together with the chain maps $f^i : C_*^i \to C_*^{i+1}$ such that the following diagram is obtained:*

$$C_*^0 \xrightarrow{f^0} C_*^1 \xrightarrow{f^1} C_*^2 \xrightarrow{f^2} \dots$$

The filtered complex in Fig. 2.4i with the inclusion maps of its simplexes becomes a *persistence complex*. Fig. 2.4ii shows part of the persistence complex with the chain complexes expanded. The filtration index increases horizontally to the right under the chain maps f_i and the dimension decreases vertically to the bottom under the boundary operators ∂_p. Observe that the boundary of a *0-simplex* is 0 therefore all 0-simplexes (e.g., vertices, singletons) are cycles. For illustration purposes, only the filtration of complexes is considered (Fig. 2.4i) and not each individual simplex in it. The filtration index i over the *chain groups* is replaced by the l_i index which identifies the simplicial complex induced by the *level line*. The filtered complex given by the pair $\{C_*^i, f^i\}$ (Fig. 2.4ii) is called *persistence module* and is defined later. The next section presents a graphical approach for computing persistent homology.

2.5.3 Computation of persistent homology

In ref. [10], the authors provided a well-established algorithm for computing persistent homology based on the reduction of an incident matrix[2] of simplexes. Here, the original persistence

---

[2]An incident matrix is a matrix that shows the relationship between two dimensions of simplexes. If the first class is V and the second is E then the matrix has one row for each element of V and one column for each element of E. The entry in row v_n and column e_m is 1 if v_n and e_m are related [33].

homology algorithm for the $\mathbb{Z}_2$ coefficients is described (introduced in ref. [16]). This method is intuitive but more significantly efficient when a large number of complexes is involved in the computations.

Step 1. Ordering the entry of simplexes

Simplexes are grouped into two categories: positive and negative.

Definition 2.27 (Positive simplex) σ *is a positive simplex if it creates a cycle when entering the filtration.*

Definition 2.28 (Negative simplex) σ *is a negative simplex if it destroys a cycle when entering the filtration.*

The category of simplexes is recorded as they appear in the filtration; a tabular data structure is created as follows (Table 2.2):

1. Simplicial complexes are enumerated according to their order of appearance.

2. The row σ is filled with the label of each simplex in the filtration.

3. The row l_i is filled with the level line in which the respective simplex in row σ appears.

4. In row sgn, the category of the simplex is annotated. Positive simplexes are denoted by "+" and negative simplexes are denoted by "-".

5. In row K_i, the entry instant of a simplex is recorded using the sequence order of step 1.

Intervals without changes are not recorded but appear as jumps in the l_i row. An arbitrary order is given to simplexes that enter at the same l_i. For example, in Table 2.2 at l_6 it is arbitrarily decided to introduce cycle "e" first than "bd". Algorithm 6 presents the pseudocode for recording the entry of the simplexes.

Algorithm 6 Ordering the entry of simplexes

```
1: procedure ORDERINGSIMPLEXESENTRY(S)              ▷ S : Set of set of simplexes
2:     K = number of abstract simplexes in S
3:     M = matrix of K columns and rows label:{K_i,σ,l_i,sgn}
4:     SetAllRow(M,K_i) = {1,...,K}
5:     k = 0;
6:     for each s ∈ S
7:         for each σ ∈ s
8:             M(σ,k) = σ
9:             obtain the level line l where the node σ was obtained
10:            M(l_i,k) = l
11:            if σ is positive simplex then M(sgn,k) = +
12:            if σ is negative simplex then M(sgn,k) = −
13:            k = k + 1
14:        endfor
15:    endfor
16:    return M
```

Table 2.2: Example of simplex ordering for the entry of simplexes of Fig. 2.4i. When a new cycle appears (e.g., "a") then it is annotated as positive while "bd" is recorded as negative because it deletes the one 0-dimension simplex. The ellipsis stands for the remaining configurations of simplexes till the construction of a fully connected simplicial complex.

l_i	0	2	3	4	6	6	8	9	9	9	9	9	9	9	9	...
σ	a	b	c	d	e	bd	ce	ed	cd	cb	eb	ea	ca	da	ab	...
sgn	+	+	+	+	+	-	-	-	+	+	-	-	+	-	+	...
K_i	0	1	2	3	4	5	6	7	8	9	10	11	12	13	14	...

Step 2. Relate cycles with boundaries

In order to measure the life-time of a non-binding cycle, its homology class is tracked (since it is created by a positive simplex) until its disappearance by a negative simplex. The row K_i marks the simplex entry. To record disappearance, a new row is added which is denoted by D_{p-1}. The sub-index $(p-1)$ is a reference to the homology dimension of the boundary elements:

1. One negative chain is selected from the table.

2. The boundary elements are obtained.

3. The boundary element with the largest filtration index is selected (given by K_i).

4. For the boundary element obtained at step 3, the value K_i of the negative chain selected at step 1 is recorded as D_{p-1}.

5. The procedure is repeated until all negative chains are processed.

For example, in Table 2.3 the negative chain bd appears at $K_i = 5$; its boundary elements are b and d and b appears at $K_i = 1$ and d at $K_i = 3$ (simplex d is the youngest). Under the row D_{p-1}, the value 5 is recorded for simplex d; thus d is created at $K_i = 3$ and is destroyed at $K_i = 5$. Algorithm 7 describes the procedure. In Table 2.2, the simplices $\{cd, cb, ca, ab\}$ are recorded as positive but in Table 2.3 they are regarded as negative due to the arbitrary order assigned to the simplexes entering the filtration. In the homology of connected components they destroy a connected component, but in the homology of simplexes of dimension one they create new loops; thus they are positive simplexes.

Table 2.3: Example of life-time records for the simplexes of Fig. 2.4i. The ellipsis stands for the remaining configurations of simplexes till the construction of a fully connected simplicial complex.

l_i	0	2	3	4	6	6	8	9	9	9	9	9	9	9	...
σ	a	b	c	d	e	bd	ce	ed	cd	cb	eb	ea	ca	da	ab	...
sgn	+	+	+	+	+	-	-	-	+	+	-	-	+	-	+	...
K_i	0	1	2	3	4	5	6	7	8	9	10	11	12	13	14	...
D_{p-1}	∞	14	12	5	6				...	...			...			...

Algorithm 7 Persistence of cycles and boundaries

```
 1:  procedure PERSISTENCE(M)                              ▷ M : Matrix of simplexes
 2:      AddOneRow(M, D_p-1)                    ▷ D_p-1 homology class disappearance index
 3:      SetAllRow(M, D_p-1) = ∞
 4:      for each column c ∈ M
 5:          if M(sgn,c) is − then                                      ▷ negative chain
 6:              ∂ = GetBoundarySimplexes(M(σ, c))               ▷ ∂ boundary elements
 7:              ∂_newest = GetTheNewestSimplex(∂)
 8:              c_newest = GetColumnGivenTheSimplex(∂_newest)
 9:              M(D_p-1, c_newest) = c
10:          endif
11:          Persistence = Matrix of same columns than M and rows label:{σ, birth, death}
12:          SetAllRow(Persistence, σ) = GetAllRow(M, σ)
13:          SetAllRow(Persistence, birth) = GetAllRow(M, K_i)
14:          SetAllRow(Persistence, death) = GetAllRow(M, D_p-1)
15:          return Persistence
16:      endfor
```

Step 3. Homology visualization

The tabular visualization is extended to a two-dimensional representation spanned by the *index* and *persistence* axes. The *k-interval* of (σ^i, σ^j) is extended into a *k-triangle* spanned by $(i, 0), (j, 0), (i, j - i)$ in the index-persistence plane. The *k-triangle* is closed along the vertical and horizontal edges and open along the diagonal connecting $(j, 0)$ to $(i, j - i)$ as shown in Fig. 2.5i. This represents the *p-cycle* created by σ^i and destroyed by σ^j. Under this compact visual representation, $\beta_p^{i,n}$ is the number of *p-triangles* that contain *(i,n)* in the index-persistence plane; each triangle covers the region for which the cycle is non-bounding. The persistent Betti numbers are *non-increasing* along vertical lines in the index-persistence plane. The same is true for lines in the diagonal direction and for all lines between the vertical and diagonal directions.

Lemma 2.2 (Monotonicity lemma)

$$\beta_p^{l,n} \leq \beta_p^{l',n'} \text{ whenever } n' \leq n \text{ and } l \leq l' \leq l + (n - n')$$

The base of *p-triangles* (Fig. 2.5i) induces a set of simplex pairs (σ_i, σ_j) each representing a *p-cycle* for $0 \leq p \leq 2$ (*p* is the dimension). Each pair is visualized on the *index* axis by a half-open interval $[i, j)$ which is called the *k-interval* [16] (Fig. 2.5ii). These *k-intervals* provide another summary about the filtration (when a cycle is created or destroyed). As an example, consider Fig. 2.5ii and the question: what is the *Betti$_0$* number between 0 and 3 $(\beta_0^{0,3})$? At 0, just one cycle is created and three units later this cycle still exists, therefore $\beta_0^{0,3} = 1$; similarly for $\beta_0^{3,2}$. At filtration index 3 three bars exist a, b, c and one appears d. Two steps later just a, b, c persist from the initial set, hence $\beta_0^{3,2} = 3$.

Step 4. Simplification

Simplicial complexes originate from the decomposition of a digital image into regions that progressively split, therefore simplexes have structural meaning. Consider complex K; during the filtration induced by the *inclusion tree*, simplexes enter according to the particular way of traversing the tree (*growth model*). The arbitrary ordering for the entry of simplexes given by the *filtration index* might not have a corresponding set in the growth model. The idea of the simplification step is to reduce the effect of this kind of conflict. A pair (σ_i, σ_j) defines a *k-cycle* that may be visualized by a *k-triangle* (Fig. 2.5iii). Whenever σ_i occurs in a conflict then it is allowed to move to a new location. This changes the Betti numbers of the reordered filtration thus being consistent with the inclusion tree.

As an example, consider the filtration of Fig. 2.5iii; a, b both enter at time 0 but the computation requires that simplexes enter one at a time. In this step, it is allowed to move a and b to time t^0 but to keep the end at the time t^1. Similarly, c, d start at time 1 so it is allowed to move c, d at time t^1. In this example, simplex c starts at time 1 and also ends at time 1, thus only simplex d can continue. This approach to persistence homology can be studied in detail in refs. [16] and [56].

2.5.4 Persistence diagram

The *k-intervals* introduced in the previous section are defined in the literature as the *persistence barcode*. An alternative method for representing persistent vector spaces is given by the *persistence diagram* which is a two-dimensional scatter plot where each point corresponds to a persistent homology class[3] (Fig. 2.6ii). The coordinates are filtration indexes of birth and death of the related homology class. To formalize the concept, a definition is given for the *persistence module* that was mentioned briefly in section 2.26.

Definition 2.29 (Persistence module) *A persistence module $\mathcal{W}$ is a family of $\mathcal{R}$ − modules M^i together with the homomorphisms ϕ^i : $M^i \longrightarrow M^{i+1}$ [58].*

Here, $\mathcal{R}$ refers to an arbitrary commutative ring with unity (def. 1.19) $\mathcal{R} = \mathbb{Z}/2\mathbb{Z}$ (see section 1.28). The homology of a persistence complex is a *persistence module* where ϕ_i maps a homology class to the one that contains it. Given the filtration index set $A \in \mathbb{R}$ then a persistent module of homology classes $\mathcal{W}_*^A$ (* is a generic placeholder for dimension) is comprised of a family of modules $C_*^{\alpha \in A} \in \mathbb{Z}/2\mathbb{Z}$ of vector spaces together with a family of linear maps $\{f^{\alpha, \beta} : C_*^\alpha \to C_*^\beta\}_{\alpha \leq \beta \in A}$ (Fig. 2.4ii), such that $\alpha \leq \beta \leq \gamma$ implying $f^{\alpha, \gamma} = f^{\beta, \gamma} \circ f^{\alpha, \beta}$ [8].

Definition 2.30 (Regular value of persistent module) *A real number α is said to be a regular value of the persistent module $\mathcal{W}_*$*

---

[3]A homology class is an equivalence class of a finite linear combination of objects with zero boundaries. Such a linear combination is considered to be homologous to zero if it is the boundary of something having dimension greater than one. For instance, two curves represent the same homology class if both form the boundary of some region.

if some $\epsilon > 0$ exists such that for all $\delta < \epsilon$ the maps $f^{\alpha-\delta,\alpha+\delta}$ are all isomorphisms. Otherwise, α is a critical value.

Definition 2.31 (Tame persistent module) *A persistent module $\mathcal{W}_*$ is called tame if it has a finite number of critical values and if all $C_*^i \in \mathbb{Z}/2\mathbb{Z}$ vector spaces are of finite rank.*

If the persistent module $\mathcal{W}_*^{0 \leq a \in A}$ is *tame* then it has the smallest non-zero critical value $\rho(\mathcal{W}_*)$ which is called the *feature size* of the persistence module. Given a tame persistent module $\mathcal{W}_*$ and a finite ordered list of critical values $0 = \xi_0 < \xi_1 < \ldots < \xi_m$, then regular values $\{a_i\}_{i=0}^m$ are chosen such that $\xi_{i-1} < a_{i-1} < \xi_i < a_i$ for all $1 \leq i \leq m$ (for short, $C_*^i \equiv C_*^{a_i}$ and $f^{i,j} : C_*^i \to C_*^j$ for $0 \leq i \leq j \leq m$).

A vector $v \in \mathcal{W}_*^{i \in A}$ is created at level i if $v \notin imf^{i-1,i}$ (where im is the image); in other words if v is not mapped from i-1 into i then v does not exist in i-1 and it is just created. Such vector is destroyed at level j if $f^{i,j}(v) \in imf^{i-1,j}$ but $f^{i,j-1}(v) \notin imf^{i-1,j-1}$ meaning that a vector is created in i and destroyed at j if in this level the image of the chain map f over v is not independent, but v and its successive images are independent during the interval i and j-1 [15]. Then $P^{i,j}$ is the vector space of vectors that are created at level i and destroyed at level j with $\beta^{i,j}$ denoting the rank (Fig. 2.6i). The information contained within a *tame* module $\mathcal{W}$ can be compactly represented by a persistence diagram.

Definition 2.32 (Persistence diagram) *This is a set of points in the upper half-plane $\{(b,d) \in \mathbb{R}^2 \mid d \geq b\}$ along with infinitely many copies of the points on the diagonal $diag = \{(x,x) \in \mathbb{R}^2\}$ denoted by $Dgm(\mathcal{W}_*)$. For each class v which is created at $\mathcal{W}_*^i$ and destroyed at $\mathcal{W}_*^j$, a point at (i,j) is drawn.*

The persistence of v is expressed by the difference between the value of death and birth: $pers(v) = j-i$. Births are plotted along the horizontal axis and deaths in the vertical axis. Since deaths happen only after births all points lie above the diagonal. The points on the diagonal correspond to trivial homology classes that are created and destroyed at every level. This provides the notion of noise; a point in the persistence diagram that is close to the diagonal represents a class that is created and destroyed fast. On the other hand, a point that is far from the diagonal has a longer life.

For application purposes, a class with points far from the diagonal represents properties that are relevant for the topology of this space.

2.6 COMPUTATIONAL COST

One approach for estimating the computational cost of this method is to characterize the relationship between the size of an image and the computing time required to obtain a persistence diagram from it. For this purpose, a set of 12 groups of images is assessed. Each group contains 40 images ranging from 50^2 to 2000^2 pixels (Fig. 2.7i):

- P-noise: synthetic images of Perlin noise rescaled and added into itself for generating fractal noise.

- LPS: images acquired from histological tissue sections of liver used as a case study.

- PNE-1, PNE-2, PNE-3: images acquired from histological tissue sections of pancreatic neuroendocrine neoplasms with faint stain intensity, average stain intensity, and histological artifacts, respectively.

- IHC2X, IHC2X2, IHC2X3: images acquired from histological tissue sections of breast cancer (HER2, 2+).

- IHC3X, IHC3X2, IHC3X3, IHC3X4: images acquired from histological tissue sections of breast cancer (HER2, 3+).

The three groups of images IHC2X, IHC2X2, and IHC2X3 belong to different histological specimens. Similarly, for the groups IHC3X, IHC3X2, IHC3X3, and IHC3X4. Results are shown in Fig. 2.7ii; they are normalized and fitted to a curve by polynomial (quadratic) regression with a coefficient of determination $R^2 > 0.9999$. In the case of the synthetic images (P-noise), linear behaviour is observed with an $R^2 > 0.995$. The processing time required to run as a function of input size is shown in Fig. 2.7iii. The results indicate that time complexity has a quasilinear behavior which is $\mathcal{O}(n^{1+\varepsilon})$ for $\varepsilon < 1$. The time cost of the algorithm is lower-bounded by an almost linear behavior obtained by synthetic images and upper-bounded by $\mathcal{O}(n^2)$ in the worst-case scenario. The image dismantling step at the beginning of the workflow introduced a massive reduction in the amount of simplicial complexes generated by the image, compared with the original method of using pixels as the vertices of simplicial complexes. The use of abstract simplicial complexes introduced a major improvement in computational efficiency in comparison with the nominal cost of computing persistent homology in a metric embedding (e.g., using Rips, Witness, or Čech complexes) which is upper-bounded by $\mathcal{O}(n^3)$ [57].

2.7   CHAPTER FIGURES

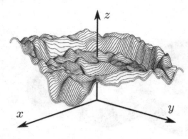

(i) Example of a 2D grey-level image defined by a function ϕ obtaining values from X (*2-dimensional* digital plane).

(ii) Example of a point $p \in X$ where X is a grey-level image and $\varepsilon = 1$. The neighborhood of a point N_p consists only of the point p. The grey region surrounding p represents points along the boundary of p.

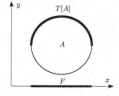

(iii) Example of a *top surface* of a 2D shape [44].

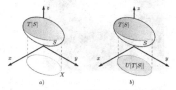

(iv) Example of a) a surface S, its **top** $T[S]$, as well as the spatial domain X induced by the b) **umbra** of the top surface.

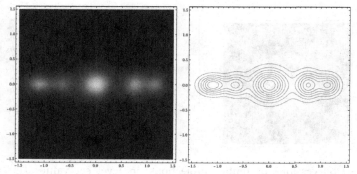

(v) Contour or level lines (right) obtained from a grey-level image (left) viewed as a topographical surface.

Figure 2.1: 2D grey-level images.

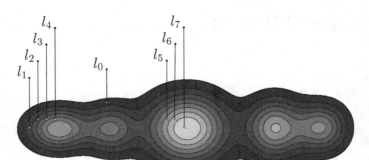

(i) An ordered set of level lines.

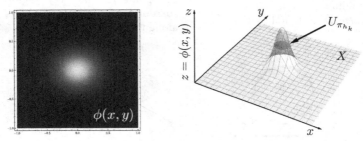

(ii) Example of $U_{\pi_{h_k}}[\phi]$ on a plane passing at level set h_k of $\phi(x,y)$. A grey-level image (left) and a 3D Cartesian system representing the image (right) are shown. X is the spatial domain of the image and $z = \phi(x,y)$ represents the digital image function; observe that the resulting projection lies in 3D space.

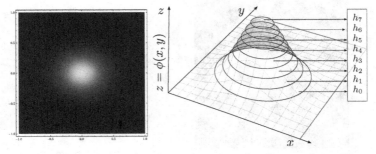

(iii) Successions of umbra projections on level set values $H_F = \{h_0, h_1, h_2, h_3, h_4, h_5, h_6, h_7\}$. A grey-level image (left) and a set of projected umbras to eight level set positions (right) are shown.

Figure 2.2: Decomposition of digital images.

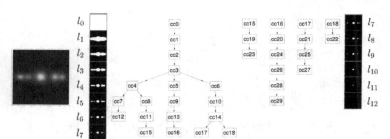

(i) An inclusion tree spanned by a hierarchical set of projected umbras. A grey-level image (left) and the tree spanned by nesting connected components of projected umbras (right) are shown. The height of the slicing plane from l_0 to l_{13} induces a progressive splitting and vanishing of connected components from one connected component covering the spatial domain at l_0 to a single component at l_{12}. For establishing an order of relations between connected components, each one of them has to be individually identified.

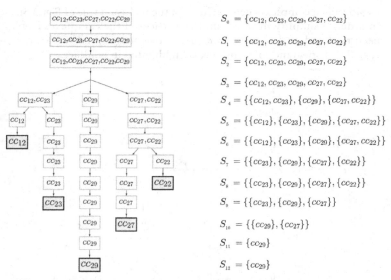

(ii) Sets obtained by considering leaf nodes $\{cc_{12}, cc_{23}, cc_{29}, cc_{27}, cc_{22}\}$ as representatives of their branch; hence a bifurcation becomes the union for two or more leaf nodes. The root of the tree is the union of all leaf nodes; this is consistent with the inclusion relations of connected components.

Figure 2.3: Inclusion trees.

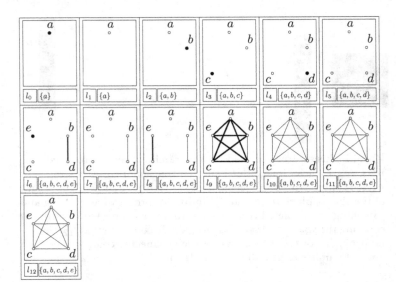

(i) Simplicial complexes generated from the inclusion tree of Fig. 2.3i. In order to simplify the diagram, the vertices of the complexes have been renamed in this way: $a = cc_{29}$, $b = cc_{27}$, $c = cc_{23}$, $d = cc_{22}$, $e = cc_{12}$. The appearance of a simplex is highlighted in boldface.

$$
\begin{array}{ccccccccc}
\partial_3 \downarrow & & \partial_3 \downarrow & & \partial_3 \downarrow & & \partial_3 \downarrow & & \partial_3 \downarrow \\
C_2^{l_{12}} & \xrightarrow{f^0} & C_2^{l_{11}} & \xrightarrow{f^1} & C_2^{l_{10}} & \xrightarrow{f^2} & C_2^{l_9} & \cdots \xrightarrow{f^{12}} & C_2^{l_0} \\
\partial_2 \downarrow & & \partial_2 \downarrow & & \partial_2 \downarrow & & \partial_2 \downarrow & & \partial_2 \downarrow \\
C_1^{l_{12}} & \xrightarrow{f^0} & C_1^{l_{11}} & \xrightarrow{f^1} & C_1^{l_{10}} & \xrightarrow{f^2} & C_1^{l_9} & \cdots \xrightarrow{f^{12}} & C_1^{l_0} \\
\partial_1 \downarrow & & \partial_1 \downarrow & & \partial_1 \downarrow & & \partial_1 \downarrow & & \partial_1 \downarrow \\
C_0^{l_{12}} & \xrightarrow{f^0} & C_0^{l_{11}} & \xrightarrow{f^1} & C_0^{l_{10}} & \xrightarrow{f^2} & C_0^{l_9} & \cdots \xrightarrow{f^{12}} & C_0^{l_0} \\
\partial_0 \downarrow & & \partial_0 \downarrow & & \partial_0 \downarrow & & \partial_0 \downarrow & & \partial_0 \downarrow \\
0 & & 0 & & 0 & & 0 & & 0
\end{array}
$$

(ii) Example of an expanded persistence complex obtained by taking the sequence of simplexes of Fig. 2.4i as a filtered complex. The filtration index increases horizontally and the dimension decreases vertically to the bottom under the boundary operators ∂_i.

Figure 2.4: Abstract simplicial complexes.

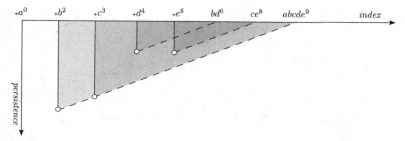

(i) Visualization of the pairing among cycles with boundaries (Table 2.3). The p-triangle of "a" is unbounded and not shown; this is a triangle covering all other triangles. The superscript shows the filtration index given by row l_i.

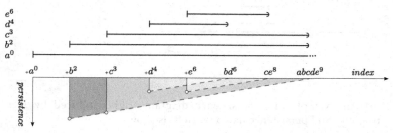

(ii) The *k-intervals* induced from the *k-triangles* in Fig. 2.5i.

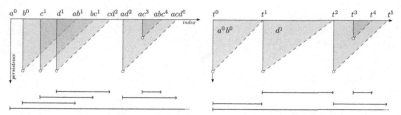

(iii) Example of the simplification step: a persistence configuration (left) and its simplification (right) are shown. Simplexes that start at the same time are joined together.

Figure 2.5: Persistent homology.

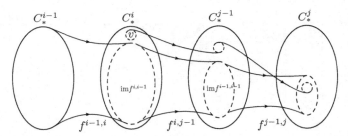

(i) The class v is created at i since it does not lie in the image of C_*^{i-1}. Furthermore, v is destroyed entering C_*^j since this is the first time its image merges into C_*^{i-1}.

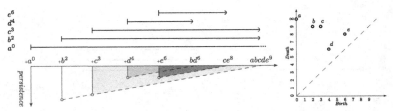

(ii) An example of a persistence diagram (right) induced by k-triangles and persistence barcodes (left) is shown.

Figure 2.6: Homology groups.

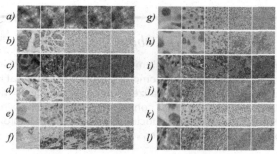

(i) Dataset for estimating computational complexity; six sample images for each group: (a) P-Noise, (b) IHC2X2, (c) IHC3X, (d) IHC2X, (e) PNE-1, (f) IHC3X2, (g) LPS, (h) PNE-3, (i) IHC3X3, (j) IHC3X4, (k) PNE-2, and (l) IHC2X3.

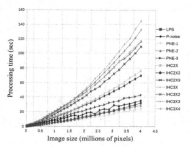

(ii) Processing time measured in sec vs. image size measured in millions of pixels.

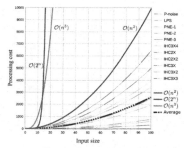

(iii) Processing time as a function of input size; curves for exponential ($\mathcal{O}(2^n)$), cubical ($\mathcal{O}(n^3)$), and quadratic ($\mathcal{O}(n^2)$) time algorithms are included as well, plus the average performance of the proposed approach.

Figure 2.7: Computational cost using synthetic and real images.

Think about it...

In order to tackle the problem of computational complexity involved in computing homology groups from digital images, an approach was realized from a sequence of morphological operations that constructed a finite approximation of the image, thus reducing the number of simplexes involved in the computations.

FURTHER READING

Caselles, V. and Monasse, P. (2009). *Geometric Description of Images as Topographic Maps.* Springer.

II

Case studies

CHAPTER 3

Recognizing noise

The chapter explores how topological features and particularly the persistent diagram recognizes noise when it is applied to digital images. For this purpose, noise generated by stochastic processes is added to synthetic images.

3.1 SYNTHETIC IMAGES

The process of generating synthetic images consists of the following steps:

1. Random generation of the 2D position of 300 points; the points are located in a circular region (508^2 pixels).

2. A single 2D isotropic (i.e., circular symmetric) Gaussian distribution with mean $\mu = (0,0)$ and standard deviation $\sigma = \{\sigma_1, \sigma_2, \dots, \sigma_8\}$ is obtained: $G(x,y) = e^{-(x^2+y^2)/(2\sigma^2)}$.

3. The cloud is normalized between $[0,1]$ with $G_{spot}(x,y) = G(x,y)/\sum_x \sum_y G(x,y)$ and then scaled to three intensity peak levels $I = \{I_1, I_2, I_3\}$ with $G_{spot}(x,y) = G_{spot}(x,y) \times I$.

4. $G_{spot}(x,y)$ is added to a blank image at the position obtained during the first step.

Three kinds of backgrounds are generated:

- A flat background image denoted by Bg is generated.

- A white noise background image denoted by Bg_w is obtained by applying white noise with $\mu = 0.05$ and $\sigma = 0.001$ to

Bg; this noise does not represent photons hitting the camera sensor but noise deriving from color saturation (Fig. 3.1i).

- A background denoted by Bg_l is obtained, simulating illumination problems, by adding in a multiplicative manner a distortion function $cos(\sqrt{(x^2 + y^2)})$ with $x = [-\pi/2 \dots \pi/2]$, $y = [-\pi/2 \dots \pi/2]$ applied on the luminance channel of Bg_w (Fig. 3.1ii).

By combining these parameters, 144 synthetic images are created. The effect of noise is shown in Fig. 3.1iii thus illustrating how difficult it is to segment the Gaussian spots in noisy environments; the intensity of the spots is in many cases less than the noise itself.

3.2 WORKFLOW

As outlined in section 2.4.1, the intensity can be defined by $I(X) : (R(X), G(X), B(X)) \longrightarrow \mathbb{R}$ where X is the spatial domain of the image and $R(X), G(X), B(X)$ is the intensity function of the three color channels (red, green, and blue). The nominal rank of I is $[0 \dots 1] \in \mathbb{R}$. The rank of I is split into κ number of segments:

$$A = \{a_i \mid a_i \in rank\,I, \ \forall i \in [0 \dots (\kappa - 1)]\ a_{i+1} - a_i = \epsilon\}$$

The number of splits $a_i \in A$ is defined a priori because the rank of I is known. Arbitrarily, $\kappa \in \mathbb{N}$ is assigned the value 1000. However, images do not distribute their intensities across all a_i values. Hence, for each image only the a_i values that the image have intensity values in them are considered. A new empirical ordered set is then defined from h_0 at the minimum a_i to h_m which is the maximal segment of intensity a_i that is present in the images; this set of real values is defined by H_F. Thus for each one of the images, a collection of projected umbras is obtained by traversing across all values of H_F (Fig. 3.1a-3.1f).

From the projected umbras, an inclusion tree (outlined in section 2.4.3) is obtained. From each inclusion tree, abstract simplicial complexes are obtained as described in section 2.4.5. The persistent homology for the simplicial chains of dimension 0 is then computed; this is the persistence of connected components in the space of simplexes. For this purpose, the inclusion tree is traversed in reverse order and the moment of appearance and merging of simplexes is recorded (outlined in section 2.5.2).

3.3   RESULTS

The points in the persistence diagram are located in a two-dimensional Cartesian plane. Let's recall that these points are in a one-to-one correspondence with the connected components of the abstract simplexes induced from the projected umbras. The birth and death times in the persistence barcode becomes the x- and y-coordinates of the persistence diagram. The points over the diagonal correspond to connected abstract simplexes that appear and disappear simultaneously; these are components that appear at h_i but disappear at h_{i+1}. Significant components are far from the diagonal and more likely represent useful features in the image. It is expected to observe the homology classes of dimension 0 for the noise and background signal to be close to the diagonal given their lack of structure. On the contrary, it is expected to observe the homology classes of dimension 0 for the Gaussian spots to be well differentiated and away from the diagonal.

In order to quantify the discriminative power of noise, a density clustering approach for topological features is applied on the persistent diagram. Density clustering allows to identify the spatial organization of a point cloud without any prior knowledge about data distribution and in particular with no prior information about the number of clusters. A rigorous introduction to density clustering is provided in ref. [27]. The clustering step is utilized in order to define a persistence interval for the larger cluster with the smaller persistence, likely noise.

Fig. 3.1iv shows four persistent diagrams obtained from four representative images of the sample set. The top left image has 300 spots with $\sigma = 0.75$, $I = 0.5$, and a Bg_l background profile; its persistent diagram demonstrates a clear differentiation between signal and noise. The top right image shows similar results but with $I = 1$ and Bg_w as the background profile; the homology classes related to spots group together and cluster in a natural way. The image on the bottom left corresponds to $\sigma = 7$, $I = 0.5$, and a Bg_l background profile; its persistent diagram shows a clear differentiation for noise. Finally, the bottom right diagram shows an image generated with $\sigma = 0.25$, $I = 0.25$, and a Bg_w background profile; even in these unfavourable settings where spots are barely distinguishable from background, the persistent diagram shows a group of spots with larger persistence (far from the diagonal). It is also interesting to observe how the shape of noise changes across the images. One could expect a consistent pattern for noise generated with the same function but this appears not to be true. This is due to the fact that Gaussian spots influence the way of obtaining the inclusion tree; larger spots correspond to more level lines appearing during the image dismantling step.

The final evaluation is carried out using the known ground-truth positions and shapes of the Gaussian spots. A normalized confusion matrix is presented in order to avoid problems caused by the large amount of background pixels in comparison to spots (Table 3.1).

Table 3.1: Classification performance of the proposed approach. Overall, good performance is demonstrated for the sample set. The expected generalization problem in images with small spots σ_1, σ_2 and large overlapping spots σ_8 is observed (TP: true positive, TN: true negative, FP: false positive, FN: false negative, FPR: false positive rate).

	Recall TP/(TP+FN)	Specificity TN/(TN+FP)	Precision TP/(TP+FP)	FPR FP/(FP+TN)
SET				
BACKGROUND				
Bg_o	0.88	0.99	0.99	0.00
Bg_w	0.85	0.80	0.81	0.20
Bg_l	0.86	0.84	0.84	0.16
STANDARD DEVIATION				
σ_1	0.92	**0.69**	**0.75**	0.31
σ_2	0.91	**0.75**	**0.78**	0.25
σ_3	0.90	0.81	0.83	0.19
σ_4	0.91	0.90	0.90	0.10
σ_5	0.91	0.99	0.99	0.01
σ_6	0.89	1.00	1.00	0.00
σ_7	0.80	0.96	0.96	0.04
σ_8	**0.68**	0.90	0.87	0.10
INTENSITY				
I_0	0.87	0.84	0.84	0.16
I_1	0.86	0.88	0.88	0.12
I_2	0.86	0.90	0.90	0.10

3.4 CHAPTER FIGURES

(i) Background image with white noise.　(ii) Background simulating illumination problems.

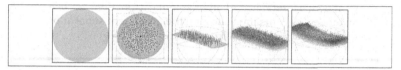

(iii) The effect of noise on synthetic images.

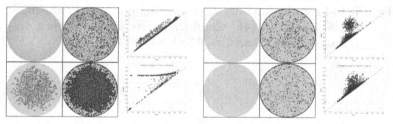

(iv) Four representative images and their persistent diagrams for homology class 0. For each image, a nontrivial segmented region is shown (in the middle); this is a region that persists for two or more filtration steps.

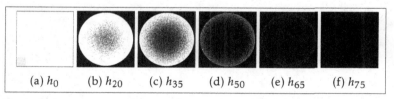

(a) h_0　(b) h_{20}　(c) h_{35}　(d) h_{50}　(e) h_{65}　(f) h_{75}

Umbra projections obtained from the σ_5 and Bg_l noise profile: six representative umbras obtained at level lines $h_0, h_{20}, h_{35}, h_{50}, h_{65}, h_{75}$.

Figure 3.1: Noise and persistent homology.

Think about it...

The section illustrates how the construction of a family of nested topological spaces, each one inducing topological features, allows a representation that summarizes the homology evolution of these spaces. The persistent diagram provides a visualization of the birth and death of topological features. From a topological point of view, features with short lifetimes are informally considered *noise* while those with a longer lifetime are considered *signal* [17]. In [9] and [18], several statistical methods are proposed to construct confidence intervals and other summary functions for persistence diagrams thus allowing the consistent separation of the topological signal from noise.

FURTHER READING

Fasy, B.T., Lecci, F., Rinaldo, A., Wasserman, L., Balakrishnan, S. and Singh A. (2014). Confidence sets for persistence diagrams. *Annals of Statistics*, 42(6), 2301-2339.

CHAPTER 4

Image segmentation

The chapter presents a computational topology implementation in the context of segmenting cell nuclei from histological images. The proposed approach is carried out in the following steps:

i An external process retrieves digital images from a database. In digital pathology, these images often consist of digitized glass slides known as whole-slide images.

ii A change in the color model is carried out in order to create a feature space that captures well the properties of the region-of-interest (ROI), e.g., cell nuclei. A feature space is sought where the persistent homology of the ROIs is different from the persistent homology of other structures in the image.

iii Images are deconstructed into connected components at different scales.

iv An inclusion tree is built that stores information between connected components.

v Persistent homology is used to summarize the *Betti number dimension-0* changes when step *(iv)* occurs.

vi Standard statistical methods are used to define a confidence interval for the birth and death of homological classes. In many cases, this interval is enough to identify an ROI in the image which makes it a kind of topological signature.

    vii Segmentation masks are obtained; this is a post-processing step for transforming selected points over the persistent diagram into a binary mask.

4.1 INTRODUCTION

The segmentation of cell nuclei is an important first step towards the automated analysis of histological images. Image segmentation methods are based on the assessment of similarities or discontinuities in the signals of interest [11]. The automated segmentation of cell nuclei is a well-studied topic for which a large number of algorithms are available in the literature and newer methods continue to be investigated [2]. A major limitation of traditional segmentation methods such as histogram-based techniques is the under/over-segmentation of cell nuclei [53].

    In order to address this and other issues, researchers proposed a variety of techniques: automated approaches to cell detection (intensity-based, region-based, active contours/level sets, probabilistic and graphical models) are reviewed in detail in ref. [24]. For example, thresholding, region growing, and watershed can locate the cell nuclei region but problems arise when these methods segment touching and overlapping nuclei. Dealing with overlapping and clustered nuclei is still a major challenge in the field. In order to tackle the problem, a variety of schemes are employed based on concavity point detection [28], distance transform [49], marker-controlled watershed [7], adaptive active contour models with shape and curvature information [3], Gaussian mixture models and expectation maximization [26], and graphs [2]. In addition, machine learning techniques [30] like Bayesian [4], support vector machines [47], AdaBoost [48], and deep learning via multiscale convolutional networks [42] are implemented as well.

    Existing methods may still have issues related to under/oversegmentation, some limitations in convergence, an increased dependency on large amounts of training data, or require explicit prior knowledge of the image structure. Due to the high variability and complexity of the color/textural content of histological images, reliable cell nuclei segmentation is still a challenging task [50]. The case study presents a computational topology approach for segmenting hepatocyte nuclei from whole-slide images.

4.2 MATERIALS AND IMAGE ACQUISITION

Liver pieces are fixed in formalin and paraffin blocks. Four μm sections are stained using the rabbit-anti-phospho-Stat3 (Tyr705) (D3A7) (1:50) (Cell Signaling Technology, USA) and the rabbit-HRP conjugated (1:100) (Santa Cruz Biotechnology, Germany) antibodies. Binding of the antibodies is visualized by the addition of the DAB substrate. Sections are counter-stained with hematoxylin.

Glass slides are imaged at high resolution in brightfield mode at a 40-fold magnification (0.23 μm/pixel) using the Hamamatsu NanoZoomer 2.0-HT (Hamamatsu Photonics, Japan). Flat field correction is done with an empty blank slide while misalignments from the line scan are corrected with a calibration slide [29]. The scanner automatically detects the regions that contain the stained tissue and also determines a valid focal plane for scanning [46]. The time requirements for scanning are around 2-4 min per slide.

Whole tissue sections are cropped into square images (616^2 pixels) which include an overlap of 20 pixels between adjacent images. The total number of images to be processed is 856. The histological landscape consists of cells with big round nuclei (hepatocytes) and cells with smaller irregularly-shaped nuclei regarded as non-parenchymal cells (hepatic stellate cells, liver sinusoidal endothelial cells, fibroblasts, endothelial cells, and Kupffer cells) (Fig. 4.1i).

4.3 FEATURE SPACE

A change in the color model is carried out in order to create a feature space that captures well the properties of the cell nuclei. The staining in the cell nuclei encodes its information in the saturation of the blue color. A grey-level image using the saturation channel S_{ch} is obtained; this is represented by the distance between the color coordinates and the luminance axis. The saturation channel is a function defined by $S_{ch}(X)$: $(R(X), G(X), B(X)) \longrightarrow \mathbb{R}$ where X is the spatial domain of the image and $R(X)$, $G(X)$, $B(X)$ are digital image functions for the red, green, and blue channels, respectively [6]. Hence, the process converts the original color image $(X, R(X), G(X), B(X))$ into a grey-level image $(X, S_{ch}(X))$. The range of S_{ch} lies in $[0, 1] \subset \mathbb{R}$.

4.4 IMAGE DISMANTLING AND INCLUSION TREE

In order to obtain a topographic map of the image, a strictly ordered set of real numbers $H_F \in \mathbb{R}$ is defined (section 2.4.1). For this purpose, the rank of S_{ch} is split into κ number of segments.

$$A = \{a_i \mid a_i \in rank\, S_{ch}, \ \forall i \in [0 \dots (\kappa - 1)] \ a_{i+1} - a_i = \epsilon\}$$

The number of splits $a_i \in A$ is known because the rank of S_{ch} is known. Arbitrarily, $\kappa \in \mathbb{N}$ is assigned the value 100. However, images do not distribute their intensities across all a_i values. Hence, for each image only the a_i values that the image have intensity values in them are considered. A new empirical ordered set is then defined from h_0 at the minimum a_i to h_m which is the maximal segment of intensity a_i that is present in the images; this set of real values is defined by H_F. For each image, a collection of umbra projections is obtained from all values of H_F. The output of this process is an inclusion tree that abstract simplicial complexes are obtained from (section 2.4.5).

4.5 FILTRATION

In order to compute the persistent homology of the 0-chains (persistence of connected components in the space of abstract simplexes), the inclusion tree is traversed in reverse order and the moment of appearance and merging of simplexes is recorded (section 2.5.2).

4.6 HOMOLOGY AND PERSISTENCE DIAGRAM

A persistence diagram (section 2.5.4) is a set of points in the upper half-plane $\{(b, d) \in \mathbb{R}^2 \mid d \geq b\}$ along with infinitely many copies of the points on the diagonal: $diag = \{(x, x) \in \mathbb{R}^2\}$. Since the 0-dimensional Betti number β_0 provides information on the number of connected components, the focus is on the persistent homology of β_0 (mainly for stability reasons). For the evolution of β_0, the instances are exactly the leaf nodes of the tree. An instance ceases to exist when it merges with another instance. For each instance v that is created at level i and ceases to exist at level j of the inclusion tree, a point at (i, j) is drawn in the persistent diagram. Points over the persistent diagrams are in a one-to-one correspondence with the connected components of the abstract simplices.

The persistence of a point v is expressed by the difference between the two values of birth and death: $persistence(v) = j - i$. Since deaths occur only after births, all points lie above the diagonal. The points on the diagonal correspond to trivial homology classes that have births and deaths at every level (the notion of noise). Thus, a point in the persistence diagram that is close to the diagonal represents a class that has a rapid birth and death. On the other hand, a point that is far from the diagonal persists longer.

4.7 STATISTICAL INFERENCE

Stained cell nuclei define regions with common textural and color properties, thus during image dismantling these regions and their constituent parts appear at several level lines of the inclusion tree (expressing longer persistence). The challenge is to define a threshold of significance (in the persistent diagram) in order to differentiate topological significant features (ROIs) from trivial ones (background). In other words, a confidence interval for the topological noise has to be computed such that the cell nuclei regions are well-defined and segmented.

The aim of statistical inference is to find such confidence interval for the distribution of persistence homology classes with unknown mean and standard deviation. The standard deviation is replaced by the estimated standard deviation s (also known as standard error). The confidence interval obtained would be exact only when the distribution of persistence (for the homology classes) is normal. Under other population distributions, the interval will be approximately correct by the central limit theorem. Since the standard error is an estimate of the true value of the standard deviation, the distribution of the sample mean follows the t-distribution with mean μ and standard deviation $s/\sqrt{n}$. The t-distribution has n degrees of freedom when the population size is $n+1$ (notation is $t(n)$). As n increases, the t-distribution becomes closer to the normal distribution since standard error approaches the true standard deviation for large n. Given a persistence diagram of n observed persistence classes x with an unknown underlying mean and standard deviation, the confidence interval for the persistence mean based on a random sample is computed by:

$$\bar{x} \pm t * s/\sqrt{n} \tag{4.1}$$

where t is the upper critical value $(1 - \delta)/2$ for the t-distribution with $n - 1$ degrees of freedom. The value for the confidence interval is estimated using an image from the set. Due to the properties of topological persistence, generalization performance is not affected.

4.8   MASK SEGMENTATION

Persistent diagrams provide clues about the kind of structures that are important. However, persistent diagrams do not carry information about the geometry of the ROIs. In order to use topological information for segmenting the cell nuclei, a post-processing operation is performed to transform the points in the persistent diagram into a properly segmented region. Region growing [1] provides a simple way to generate precise segmentations that are initialized by the proposed approach. This initialization is robust against color distortions and image blur.

4.9   VALIDATION WITH CASE STUDY DATA

An automated segmentation pipeline validated by experts is used to measure the performance of the proposed approach. The validation set consists of binary masks obtained by standard methods such as background removal, stain color deconvolution, and post-processing operations.

2D histogram variance thresholding separates regions of interest (tissue) and background pixels. Background removal is beneficial for the next step which involves stain separation using the widely-used technique of color deconvolution; a methodology that allows to separate image components based on color parameters. Virtually every set of three colors can be separated even in the case of co-localization of stains [36]. Images are reconstructed for each stain separately, requiring only knowledge of the RGB color vectors of each stain. Hence, the contribution of each stain is computed and images are produced. These grayscale images are then binarized directly by replacing all pixels with luminance >0.5 with one, and all other pixels with zero. Finally, post-segmentation operations refined image characteristics, eliminated redundancies, and improved object perceptibility (i.e., filtering, morphological operations, and watershed transform) [46]. Images went through manual quality control to ensure that desired image features are detected in an efficient manner.

4.10   VALIDATION WITH ILASTIK

Further validations are carried out using the software tool ilastik v.1.1.8 [41]. Ilastik is a general-purpose state-of-the-art image segmentation tool that provides an automated workflow based on the supervised training of a random forest classifier. For comparison purposes, training is performed manually on the same two images used to tune the proposed approach. The software is trained using color, edge, and textural features.

4.11   RESULTS WITH CASE STUDY DATA

The validation set is provided in two groups: (i) all segmented nuclei and (ii) positive cell nuclei (hepatocytes). The output mask is compared with the validation mask which contains all segmented cell nuclei: hepatocyte and non-parenchymal cells (accuracy: 84.6%) (Fig. 4.1ii). In addition, the output mask is compared with the validation mask containing hepatocyte cell nuclei only (accuracy: 86.2%) (Fig. 4.1iii). To perform this comparison, the segmented nuclei are separated into two groups according to their eccentricities. The eccentricity is the ratio of the distance between the foci of the ellipse containing a nucleus and its major axis length (value between 0 and 1). Eccentricities equal to 0 indicate a circle, while values equal to 1 indicate a line segment. A *positive* label is assigned to cell nuclei with an eccentricity ≤ 0.9 and a *negative* label to nuclei with an eccentricity > 0.9.

4.12   RESULTS WITH ILASTIK

The corresponding ROC curve for all segmented cell nuclei comparing the proposed approach with the segmentation obtained from the software tool ilastik (accuracy: 75.5%) is shown in Fig. 4.1iv. This highlights the problem of generalization in this machine learning approach. The issue arises because training is based on appearance features and the set used for training does not contain enough variability in the appearance of cell nuclei in order to segment unseen samples successfully. Increasing the number of samples in the training set and re-training for new samples are conventional strategies to minimize these problems. In practice, it is impossible to predict the complete pool of possible variations in the appearance of cell nuclei. The proposed approach achieves better separation between classes (nuclei and background) and in consequence better segmentation performance, because it does not depend on cell nuclei appearance features but in a highly abstract representation of the image which is less sensitive to changes in color, texture, and edge information.

4.13   CHAPTER FIGURES

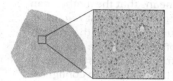

(i) Whole-slide image of liver tissue (left) and a cropped square section (1500^2 pixels) (right).

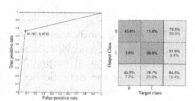

(ii) Average segmentation performance for all cell nuclei (hepatocyte and non-parenchymal cells). Normalization balances background (0) and nuclei class (1); each class contains approx. 50% of the data. Overall accuracy: 84.6%; error rate: 15.4%; specificity: 92.8%; true positive rate for nuclei: 76.7%; false positive rate: 7.2%.

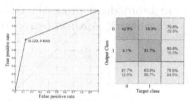

(iv) Average segmentation performance obtained with ilastik for all cell nuclei (hepatocyte and non-parenchymal cells). Ilastik is trained with the same two images used for tuning the proposed approach. Normalization balances background (0) and nuclei class (1); each class contains approx. 50% of the data. Overall accuracy: 75.5%; error rate: 24.5%; specificity: 87.7%; true positive rate for nuclei: 63.3%; false positive rate: 12.3%.

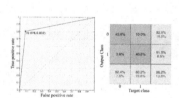

(iii) Average segmentation performance for hepatocyte cell nuclei. Normalization balances background (0) and nuclei class (1); each class contains approx. 50% of the data. Overall accuracy: 86.2%; error rate: 13.8%; specificity: 92.4%; true positive rate for nuclei: 80.2%; false positive rate: 7.6%.

Figure 4.1: Performance of image segmentation workflow.

Think about it...

The findings indicate that the proposed approach generalizes well and provides stable performance across a large set of histological images. This means that the method can generalize the definition of a region-of-interest regarding its topological features. The results also reveal that performance depends on the choice of feature space where cell nuclei are detected during the fragmentation step. This is a feature space where it is possible to generate configurations of abstract simplicial complexes.

FURTHER READING

Rojas Moraleda, R., Xiong, W., Halama, N., Breitkopf-Heinlein, K., Dooley, S., Salinas, L., Heermann, D.W., Valous, N.A. (2017). Robust detection and segmentation of cell nuclei in biomedical images based on a computational topology framework. *Medical Image Analysis*, 38, 90-103.

CHAPTER 5

Point cloud characterization

The chapter presents a computational topology implementation in the context of identifying fundamental properties of point clouds acquired from experiments. In this case, the point cloud represents the locations of chromatin markers in eukaryotic cells. The goal is to characterize chromatin structures during the healing process caused by ionizing radiation damage.

5.1 INTRODUCTION

A major goal of ongoing research in biophysics is to understand the potential interaction of nuclear architecture and function. This is useful for elucidating the mechanistic principles behind the spatial organization of chromatin and its rearrangements during intra-nuclear processes like DNA repair after exposure in ionizing radiation.

The complex repair machinery activated by chromatin damage (caused by ionizing radiation) is crucial for studying the functionally correlated chromatin architecture in the cell nuclei. These repairs are accompanied by local chromatin rearrangements and structural changes. Using super-resolution localization microscopy (spectral position determination microscopy: SPDM), it is possible to study the location of individual histone proteins in the cell nuclei. The cells are exposed to ionizing radiation of different doses and aliquots are fixed after different repair times for SPDM imaging.

Assuming that the irregularly distributed chromatin regions act as inhomogeneous Riemannian manifolds, then a Euler characteristic approach is utilized which is based on the Lipschitz-Killing curvature (LKC) of the kernel density estimator (KDE) of an unobservable underlying probability density function; this is comparable to the approach described in section 2.4.1. The LKCs of the KDE level-sets computed from the persistent homology of a point cloud can provide information on where and at which scale these nanostructures (non-random fields) reside. Statistical analysis between the persistent diagrams can also show how these structures differ depending on radiation dose and repair times.

5.2 POINT CLOUD

A point cloud is obtained from localization images; this is achieved using the approach of ref. [55]. More specifically, localization images are segmented based on a density threshold procedure to exclude regions outside the nuclear membrane as well as chromatin-depleted regions inside the nucleus (e.g., nucleoli). Fig. 5.1i shows an example of a localization image of histone H2A in the nucleus of a HeLa cell labeled with a yellow fluorescent protein. The localized fluorophores are blurred with a Gaussian function corresponding to their localization accuracy. The wide-field image of Fig. 5.1i is presented in Fig. 5.1ii. Localized histones after merging the acquired time series of SPDM images are shown in Fig.5.1iii; each localized fluorophore is represented by a white point. Fig. 5.1iv presents a histone point cloud obtained by thresholding the KDE.

5.3 WORKFLOW

Let's assume that the point cloud retrieved from the SPDM image is an inhomogeneous Riemannian manifold M. LKCs are geometric objects that depend on the Riemannian metric on M such that $\mathcal{L}_k(M)$ is a measure of the k-dimensional size of M. With a Gaussian kinematic formula, let $M \in \mathbb{R}^d$ and $D \in \mathbb{R}^k$ be well-behaved stratified spaces. If $f = (f_1, \cdots, f_k) : M \to \mathbb{R}^k$ is a C^2 k-dimensional Gaussian field (f is used as the Gaussian KDE function) satisfying the conditions above, then:

$$\mathbb{E}\{\mathcal{L}_i(f^{-1}(D))\} = \sum_{j=0}^{dimM-i} \begin{bmatrix} i+j \\ j \end{bmatrix} (2\pi)^{-\frac{j}{2}} \mathcal{L}_{i+j}(M) \mathcal{M}_j^{\gamma}(D) \qquad (5.1)$$

In this application, $k = 1, d = 2$ and only $\mathcal{L}_0$ and $\mathcal{L}_1$ are taken into account (when $i = 0$ then $\mathcal{L}_0 = \chi(M)$ is its Euler characteristic).

This form of filtration is illustrated in Fig. 5.1i resulting to the barcodes for the KDE level-sets (excursion sets). The top seven boxes (Fig. 5.1i) show the surfaces of the KDE above the excursion sets A_u for different density levels u. As the dotted lines pass through the boxes labeled H_0 and H_1, the number of intersections with bars in the $H_0(H_1)$ box provides the number of connected components in A_u. The horizontal lengths of the bars indicate how long different topological structures (generators of homology groups) persist.

5.4   RESULTS

A dataset consisting of 360 SPDM images as point clouds is used for validation. Green and yellow fluorescent protein markers are used for labelling chromatin. Antibodies for the histones H3K4 and H4K20 are used to label euchromatin and heterochromatin, respectively. The results show that heterochromatic regions alone indicate a relaxation after radiation exposure and recondensation during repair, while euchromatin seems to be unaffected or behaves in a contrary fashion (Fig. 5.1ii). This differentiation can be seen by the respective persistent diagrams (Fig. 5.1i).

5.5 CHAPTER FIGURES

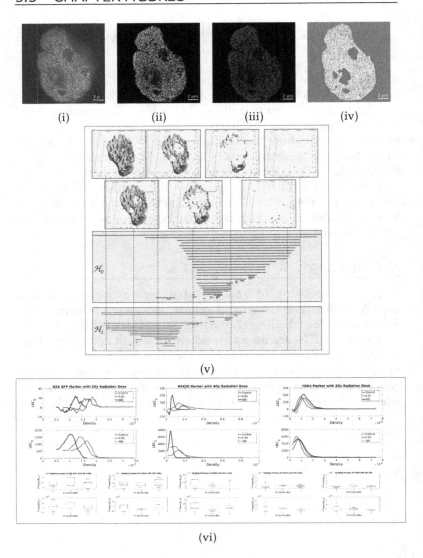

Figure 5.1: Point cloud characterization: (i) localization image of histone H2A, (ii) corresponding wide-field image, (iii) localized histones after data processing, (iv) histone point cloud, (v) KDE surfaces above the excursion sets for different density levels and barcodes indicating persistence, and (vi) results from a 360 sample point cloud dataset (RD: radiation dose, ET: time elapsed between radiation and fixation).

Think about it...

Computational topology in combination with super-resolution localization microscopy is a robust descriptor of point cloud data that correspond to structural changes in chromatin. This allows an analytic elucidation of chromatin re-arrangements after irradiation and during repair.

FURTHER READING

Bohn, M., Diesinger, P., Kaufmann, R., Weiland, Y., Müller, P., Gunkel, M., Von Ketteler, A., Lemmer, P., Hausmann, M., Heermann, D.W., and Cremer C. (2010). Localization microscopy reveals expression-dependent parameters of chromatin nanostructure. *Biophysical Journal*, 99(5), 1358-1367.

Hofmann, A., Krufczik, M., Heermann, D.W., and Hausmann, M. (2018). Using persistent homology as a new approach for super-resolution localization microscopy data analysis and classification of H2AX foci/clusters. *International Journal of Molecular Sciences*, 19(8), 2263.

Bibliography

[1] R. Adams and L. Bischof. Seeded region growing. *IEEE T. Pattern Anal.*, 16:641–647, 1994.

[2] Yousef Al-Kofahi, Wiem Lassoued, William Lee, and Badrinath Roysam. Improved automatic detection and segmentation of cell nuclei in histopathology images. *IEEE T. Bio-Med. Eng.*, 57:841–852, 2010.

[3] S. Ali and A. Madabhushi. An integrated region-, boundary-, shape-based active contour for multiple object overlap resolution in histological imagery. *IEEE T. Med. Imaging*, 31(7):1448–1460, 2012.

[4] A.N. Basavanhally, S. Ganesan, S. Agner, J.P. Monaco, M.D. Feldman, J.E. Tomaszewski, G. Bhanot, and A. Madabhushi. Computerized image-based detection and grading of lymphocytic infiltration in her2+ breast cancer histopathology. *IEEE T. Bio-Med. Eng.*, 57(3):642–653, 2010.

[5] Stéphane Bonnevay. Pretopological operators for gray-level image analysis. *Studia Informatica Universalis*, February 2008.

[6] Vicent Caselles and Pascal Monasse. *Geometric Description of Images as Topographic Maps*. Springer Publishing Company, Incorporated, 1st edition, 2009.

[7] S. Di Cataldo, E. Ficarra, A. Acquaviva, and E. Macii. Automated segmentation of tissue images for computerized IHC analysis. *Comput. Meth. Prog. Bio.*, 100(1):1–15, 2010.

[8] Frédéric Chazal, David Cohen-Steiner, Marc Glisse, Leonidas J. Guibas, and Steve Y. Oudot. Proximity of persistence modules and their diagrams. In *Proceedings of the Twenty-fifth Annual Symposium on Computational*

Geometry, SCG '09, pages 237–246, New York, NY, USA, 2009. ACM.

[9] Frédéric Chazal, Brittany Terese Fasy, Fabrizio Lecci, Bertrand Michel, Alessandro Rinaldo, and Larry A. Wasserman. Robust topological inference: Distance to a measure and kernel distance. *CoRR*, abs/1412.7197, 2014.

[10] David Cohen-Steiner, Herbert Edelsbrunner, and Dmitriy Morozov. Vines and vineyards by updating persistence in linear time. In *Proceedings of the Twenty-second Annual Symposium on Computational Geometry*, SCG '06, pages 119–126, New York, NY, USA, 2006. ACM.

[11] Domingos Lucas Latorre de Oliveira, Marcelo Zanchetta do Nascimento, Leandro Alves Neves, Moacir Fernandes de Godoy, Pedro Francisco Ferraz de Arruda, and Dalisio de Santi Neto. Unsupervised segmentation method for cuboidal cell nuclei in histological prostate images based on minimum cross entropy. *Expert Syst. Appl.*, 40:7331–7340, 2013.

[12] M.M. Deza and E. Deza. *Encyclopedia of Distances*. Springer Berlin Heidelberg, 2014.

[13] U. Eckhardt and L. Latecki. *Digital Topology*. Current Topics in Pattern Recognition Research, (Research Trends). Council of Scientific Information, Vilayil Gardens, Kaithamukku, Trivandrum, India, 1995.

[14] H. Edelsbrunner, D. Letscher, and A. Zomorodian. Topological persistence and simplification. In *Proceedings of the 41st Annual Symposium on Foundations of Computer Science*, FOCS '00, pages 454–, Washington, DC, USA, 2000. IEEE Computer Society.

[15] Herbert Edelsbrunner and John L. Harer. *Computational Topology: An Introduction*. American Mathematical Society, Providence (R.I.), 2010.

[16] Herbert Edelsbrunner, David Letscher, and Afra Zomorodian. Topological persistence and simplification. *Discrete & Computational Geometry*, 28(4):511–533, 2002.

[17] Brittany Terese Fasy, Fabrizio Lecci, Alessandro Rinaldo, Larry Wasserman, Sivaraman Balakrishnan, and Aarti Singh.

Confidence sets for persistence diagrams. *Ann. Stat.,* 42(6):2301–2339, 2014.

[18] Brittany Terese Fasy, Fabrizio Lecci, Alessandro Rinaldo, Larry Wasserman, Sivaraman Balakrishnan, and Aarti Singh. Confidence sets for persistence diagrams. *Annals of Statistics,* 2014. *ArXiv preprint ArXiv:1303.7117,* 42(6): 2301–2339, 2014.

[19] P.J. Giblin. *Graphs, Surfaces, and Homology: An Introduction to Algebraic Topology.* Chapman and Hall mathematics series. Chapman and Hall, 1977.

[20] P.R. Halmos. *Naive Set Theory.* Undergraduate Texts in Mathematics (printed since 1960), ISBN:9781475716450. Springer New York, 2013.

[21] A. Hatcher. *Algebraic Topology.* Cambridge University Press, 2002.

[22] P.J. Hilton and S. Wylie. *Homology Theory: An Introduction to Algebraic Topology.* Cambridge University Press, 1967.

[23] Danijela Horak, Slobodan Maletic, and Milan Rajkovic. Persistent homology of complex networks. *Journal of Statistical Mechanics: Theory and Experiment,* 2009(03):P03034, 2009.

[24] Humayun Irshad, Antoine Veillard, Ludovic Roux, and Daniel Racoceanu. Methods for nuclei detection, segmentation, and classification in digital histopathology: a review-current status and future potential. *IEEE Rev. Biomed. Eng.,* 7:97–114, 2014.

[25] K. Itô and Nihon-Sūgakkai. *Encyclopedic Dictionary of Mathematics.* Number Vol. 1 & vol. 2. MIT Press, Cambridge, MA, USA, 1993. Mathematical Society of Japan.

[26] Chanho Jung, Changick Kim, Seoung Wan Chae, and Sukjoong Oh. Unsupervised segmentation of overlapped nuclei using bayesian classification. *IEEE T. Bio-Med. Eng.,* 57(12):2825–2832, 2010.

[27] Rinaldo A. Verstynen T. Kent, B. P. A python package for interactive density-based clustering. *ArXiv preprint arXiv:1307.8136,* 2013.

[28] Hui Kong, M. Gurcan, and K. Belkacem-Boussaid. Partitioning histopathological images: an integrated framework for supervised color-texture segmentation and cell splitting. *IEEE T. Med. Imaging*, 30(9):1661–1677, 2011.

[29] B. Lahrmann, N. Valous, U. Eisenmann, Wentzensen, N., and Niels Grabe. Semantic focusing allows fully automated single-layer slide scanning of cervical cytology slides. *PLoS ONE*, 8(4):e61441, 04 2013.

[30] Anant Madabhushi and George Lee. Image analysis and machine learning in digital pathology: challenges and opportunities. *Med. Image Anal.*, 33:170–175, 2016.

[31] D.P. Mukherjee, N. Ray, and S.T. Acton. Level set analysis for leukocyte detection and tracking. *Image Processing, IEEE Transactions on*, 13(4):562–572, April 2004.

[32] Mark Nixon and Alberto S. Aguado. *Feature Extraction & Image Processing, Second Edition*. Academic Press, 2nd edition, 2008.

[33] Sriram Pemmaraju and Steven Skiena. *Computational Discrete Mathematics: Combinatorics and Graph Theory with Mathematica ®*. Cambridge University Press, New York, NY, USA, 2003.

[34] James F. Peters. *Topology of Digital Images: Visual Pattern Discovery in Proximity Spaces*. Springer Publishing Company, Incorporated, 2014.

[35] Vanessa Robins. *Computational Topology at Multiple Resolutions: Foundations and Applications to Fractals and Dynamics*. University of Colorado at Boulder, 2000.

[36] A. C. Ruifrok and D. A. Johnston. Quantification of histochemical staining by color deconvolution. *Analytical and quantitative cytology and histology / the International Academy of Cytology [and] American Society of Cytology*, 23(4):291–299, August 2001.

[37] Robert Sedgewick. *Algorithms in C, Parts 1-4: Fundamentals, Data Structures, Sorting, Searching (3rd Edition) (Pts. 1-4)*. Addison-Wesley Professional, 1997.

[38] J. Serra. *Image Analysis and Mathematical Morphology: Theoretical Advances*. Image Analysis and Mathematical Morphology. Academic Press, 1988.

[39] Jean Serra and Pierre Soille, editors. *Proceedings of the 2nd International Symposium on Mathematical Morphology and Its Applications to Image Processing, ISMM 1994, Fontainebleau, France, September 1994*, volume 2 of *Computational Imaging and Vision*. Kluwer, 1994.

[40] G.F. Simmons. *Introduction to Topology and Modern Analysis*. International series in pure and applied mathematics. Krieger Publishing Company, 1963.

[41] Christoph Sommer, Christoph Straehle, Ullrich Kothe, and Fred A. Hamprecht. Ilastik: interactive learning and segmentation toolkit. In *Proceedings of the IEEE International Symposium on Biomedical Imaging: From Nano to Macro, pp. 230–233*, 2011.

[42] Youyi Song, Ling Zhang, Siping Chen, Dong Ni, Baiying Lei, and Tianfu Wang. Accurate segmentation of cervical cytoplasm and nuclei based on multiscale convolutional network and graph partitioning. *IEEE T. Bio-Med. Eng.*, 62(10):2421–2433, 2015.

[43] Marta Macho Stadler. Que es la topologia, what is topology. *Revista Sigma*, Zka. 20(1), 2002.

[44] Stanley R Sternberg. Grayscale morphology. *Comput. Vision Graph. Image Process.*, 35(3):333–355, September 1986.

[45] Heinrich Tietze. Über die topologischen invarianten mehrdimensionaler mannigfaltigkeiten. *Monatshefte für Mathematik und Physik*, 19(1):1–118, dec 1908.

[46] Nektarios Valous, Bernd Lahrmann, Wei Zhou, Roland Veltkamp, and Niels Grabe. Multistage histopathological image segmentation of iba1-stained murine microglias in a focal ischemia model: Methodological workflow and expert validation. *Journal of Neuroscience Methods*, pages 250–262, 2013.

[47] Antoine Veillard, Maria S Kulikova, and Daniel Racoceanu. Cell nuclei extraction from breast cancer histopathology images using colour, texture, scale and shape information. *Diagn. Pathol.*, 8(Suppl 1):S5, 2013.

[48] J.P. Vink, M.B. Van Leeuwen, C.H.M. Van Deurzen, and G. De Haan. Efficient nucleus detector in histopathology images. *J. Microsc.*, 249(2):124–135, 2012.

[49] C. Wahlby, I.-M. Sintorn, F. Erlandsson, G. Borgefors, and E. Bengtsson. Combining intensity, edge and shape information for 2d and 3d segmentation of cell nuclei in tissue sections. *J. Microsc.*, 215(1):67–76, 2004.

[50] Pin Wang, Xianling Hu, Yongming Li, Qianqian Liu, and Xinjian Zhu. Automatic cell nuclei segmentation and classification of breast cancer histopathology images. *Signal Process.*, 122:1–13, 2016.

[51] E.W. Weisstein. *CRC Concise Encyclopedia of Mathematics, Second Edition.* CRC Press, 2002.

[52] Norman Wildberger. Set theory: Should you believe? Technical report, School of Maths UNSW, 2015.

[53] Hongming Xu, Cheng Lu, and Mrinal Mandal. An efficient technique for nuclei segmentation based on ellipse descriptor analysis and improved seed detection algorithm. *IEEE J. Biomed. Health Inform.*, 18:1729–1741, 2014.

[54] E. Zermelo. Untersuchungen Über die grundlagen der mengenlehre. i. *Mathematische Annalen*, 65(2):261–281, jun 1908.

[55] Yang Zhang, Gabriell Máté, Patrick Müller, Sabina Hillebrandt, Matthias Krufczik, Margund Bach, Rainer Kaufmann, Michael Hausmann, and Dieter W. Heermann. Radiation induced chromatin conformation changes analysed by fluorescent localization microscopy, statistical physics, and graph theory. *PLoS ONE*, 10(6):1–23, 06 2015.

[56] Afra Zomorodian and Gunnar Carlsson. Computing persistent homology. *Discrete Comput. Geom.*, 33(2):249–274, February 2005.

[57] Afra Zomorodian and Gunnar Carlsson. Computing persistent homology. *Discrete Comput. Geom.*, 33:249–274, 2005.

[58] A.J. Zomorodian. *Topology for Computing.* Cambridge Monographs on Applied and Computational Mathematics. Cambridge University Press, 2005.

Index

Printed in the United States
by Baker & Taylor Publisher Services